General Preface to the Series

Because it is no longer possible for one textbook to cover the whole field of biology while remaining sufficiently up to date, the Institute of Biology proposed this series so that teachers and students can learn about significant developments. The enthusiastic acceptance of 'Studies in Biology' shows that the books are providing authoritative views of biological topics.

The features of the series include the attention given to methods, the selected list of books for further reading and, wherever possible, suggestions for practical work.

Readers' comments will be welcomed by the Education Officer of the Institute.

1982

Institute of Biology
41 Queen's Gate
London SW7 5HU

Preface

Biologists are increasingly taught to approach problems from first principles. This is generally a good thing, but it can lead to difficulties if all the relevant information is not used. For example, when we ask: 'Is evolution true?', the discussion tends to revolve round gaps in evidence. This is the wrong centre for the debate. Evolutionary ideas form a synthesis involving virtually the entire span of biological disciplines. Someone starting from scratch is faced with a jigsaw which took an army of biologists over a century to complete, involving fact and theory, error and arrogance, superstition and success. *Neo-Darwinism* is the picture on the jigsaw box, as it were, charting the progress in our understanding of evolution from Darwin and Wallace's original paper in 1858 to the present. Most biologists tend to be impatient with history; I hope that readers of *Neo-Darwinism* will be able to bear enough of it to recognize that virtually all the criticisms about evolution raised today are repeats of old ones. Knowing the strength and solution of past debates can save time in dealing with the present situation, quite apart from forming a spring-board for the next advances in biology overall.

Professor Arthur Cain of Liverpool University and Mr Ian Lacey of Shrewsbury School read the book in draft, and made various criticisms. Most of them I have incorporated; my thanks are due to them for their comments, and my apologies for sticking to the points where I think I am right and they are wrong.

London, 1982

R.J.B.

Contents

1 Darwin, Darwinism, and Neo-Darwinism

1.1 Closet biology

The word 'biology' means, literally, the study of living things, but we all know how much easier it is to confine ourselves to dead organisms in the comfort of the laboratory or museum. This has meant that the link between the field naturalist working with the formidable complexity of real animals and plants, and the closet biologist necessarily but artificially simplifying the material at his command, is fragile, and too often breaks. (A closet biologist is one who expounds his wisdom from an indoor sanctum, basing himself on an extensive acquaintance with a restricted body of facts.) This broken link is particularly important in understanding the arguments that occur about evolution, for the simple reason that evolution is not a subject in its own right but a synthesis of disciplines as wide as biology itself: anatomy and anthropology; biometrics and biochemistry; ecology and ethology; genetics and geology; physiology and phylogeny; and so on. Few people can adequately cover this span, and, as we shall see, virtually all the criticisms about evolution since Darwin first put forward his ideas have come from genuine misunderstandings. As a matter of history, most have been raised by 'closet biologists', although at times field-workers have contributed to the general confusion (as in the gulf that opened in the 1920s between palaeontologists and other evolutionary biologists).

Five episodes of doubt about Darwinism can be recognized:
1) Objections expressed in Darwin's own time and largely anticipated in the *Origin of Species*.
2) Arguments between the biometricians and Mendelians around the turn of the century.
3) A rift between palaeontologists and geneticists in the 1920s and 1930s, which led to the general consensus usually known as the neo-Darwinian synthesis.
4) A conflict between neutralists and selectionists in the 1960s.
5) A series of independent assaults on neo-Darwinism in the 1980s by an assorted collection of palaeontologists, cladists, philosophers, and creationists.

We shall see the strengths and replies to all these in the following pages. However it is worth noting that the problems in every case have arisen because of confusion about the nature and maintenance of inherited variation. Indeed, this book is mostly about variation, and is really a cautionary parable for specialists, with the moral that they neglect data outside their own expertise at their peril.

1.2   Darwin and Wallace

Although Darwinism and evolution are often used as synonyms, they are not the same thing. Darwin himself acknowledged a galaxy of biologists who had believed before him that species were subject to change: Marchant (1719), Montesquieu (1721), Buffon (1749), Maupertuis (1754), Diderot (1764), Erasmus Darwin (1794), St. Hilaire (1795), Goethe (1795), Lamarck (1801), and Moritzi (1842). Robert Chambers in the *Vestiges of Creation* (first published 1844, tenth edition 1853) had alerted Britain to the idea of evolution maintaining that 'species are not immutable productions . . . (although, animated beings from the simplest and oldest up to the highest and most recent are under the providence of God'. Even Darwin's particular contribution of natural selection was described by Wells in a paper read to the Royal Society of London in 1813 and by Matthew in a book on *Naval Timber and Arboriculture* published in 1831.

Notwithstanding it was Charles Darwin's *The Origin of Species by means of natural selection or the preservation of favoured races in the struggle for life* (first published 1859, sixth and final edition 1872) that started the general acceptance of evolutionary ideas by both the scientific and the general world. The reason for the immediate success of the *Origin* was Darwin's explanations for the distribution of animals and plants, and his convincing interpretation of the significance of vestigial organs. Other lines of evidence, from fossils, anatomical likenesses and so on were fairly well-known to Darwin's contemporaries but were explained away before an acceptable mechanism of evolution was available. For example, the existence of fossils of extinct species was interpreted as either the remains of former creations destroyed by God as He almost destroyed the present world in Noah's flood, or divinely created artefacts, put in the rocks to confuse godless scientists. In the longer term and much more important, it was Darwin's easily-understood *mechanism* of evolution which was his most important contribution. The need for a mechanism before a scientific idea is generally accepted also occurred with continental drift, which was put forward in detail by Wegener in 1915, but not commonly accepted until the nature of tectonic plates was described by geophysicists in the 1960s.

Darwin's doubts about the immutability of species arose from his study of geographical variation in both fossils and living forms during his time as naturalist on board the survey ship *Beagle* (1831–6). For example he wrote in his *Journal* about the Galapagos Islands that 'I never dreamed that islands, about 50 or 60 miles apart, and most of them in sight of each other, formed of precisely the same rocks, placed under a quite similar climate, rising to nearly equal height, would be differently tenanted; but . . . I obtained sufficient materials to establish this most remarkable fact in the distribution of organic beings'. Back in London, Darwin began in July 1837 to make notes on the transmutation of species. In 1838 he read 'for amusement' Malthus's *Essay on the Principle of Population*, and 'being well prepared to appreciate the struggle for existence . . . it at once struck me that under these circumstances favourable variations would tend to be preserved, and unfavourable ones to be destroyed. The result of this would be the formation of new species. Here then I had at last got a theory by which to work'.

Darwin wrote down his ideas for the first time in 1842, whilst on a brief visit to his in-laws (Josiah Wedgwood II, the potter) and his parents (in Shrewsbury). He expanded his 1842 sketch in 1844, and this latter version formed the basis of his first public pronouncement on the subject in 1858.

Darwin's original intention was to write a definitive book on evolution. However in the spring of 1858 he was sent by Alfred Russel Wallace an essay 'On the tendency of varieties to depart indefinitely from the original type', written by Wallace whilst recovering from fever in the Moluccas. Darwin felt that this should be published, but on the urging of his friends Charles Lyell (author of the *Principles of Geology*, which Darwin had read on the *Beagle*, and which had first alerted him to the reality of long-continued gradual geological change) and J. D. Hooker (son of the effective founder of Kew Botanic Gardens, and instigator of the study of plant geography), he allowed a revised version of his 1844 essay to be forwarded with it to the Linnean Society.

Darwin's and Wallace's papers were read at a Linnean Society meeting on 1st July 1858, and published in the Society's *Journal* on the 20th August. They attracted little attention; the President of the Society declared in his report for 1858 that 'The year which has passed . . . has not been marked by any of those striking discoveries which revolutionise the department of science on which they bear . . . A Bacon or a Newton, an Oersted or a Wheatstone, a Davy or a Daguerre, is an occasional phenomenon, whose existence and career seem to be specially appointed by Providence, for the purpose of effecting some great important change in the condition or pursuits of man'. Controversy only broke out after the publication of the *Origin of Species* in the following year, and more vehemently after the confrontation between T. H. Huxley and 'Soapy Sam' Wilberforce, Bishop of Oxford (inadequately briefed by Richard Owen, first director of the Natural History Museum in South Kensington) at the British Association meeting in Oxford on 30th June 1860.

1.2.1 Darwin's paper to the Linnean Society

Since we are concerned in this book with the vicissitudes of the ideas which became public at the Linnean Society in 1858, it is worth recalling the actual words used. Darwin wrote 'All nature is at war, one organism with another, or with external nature. Seeing the contented face of nature, this may at first well be doubted: but reflection will inevitably prove it to be true. The war, however, is not constant, but recurrent in a slight degree at short periods, and more severely at occasional more distant periods; and hence its effects are easily overlooked . . . But the amount of food for each species must, *on an average*, be constant, whereas the increase of all organisms tends to be geometrical, and in a vast majority of cases at an enormous ratio . . . In the majority of cases it is most difficult to imagine where the check falls – though generally, no doubt, on the seeds, eggs, and young. Lighten any check in the least degree, and the geometrical powers of increase in every organism will almost instantly increase the average number of the favoured species. Nature may be compared to a surface on which rest ten thousand sharp wedges touching each other and driven inwards by incessant blows. Fully to realize these views much reflection is requisite'.

'But let the external conditions of a country alter . . . can it be doubted from the struggle each individual has to obtain subsistence, that any minute variation in structure, habits, or instincts, adapting that individual better to the new conditions, would tell upon its vigour and health? In the struggle it would have a better *chance* of surviving; and those of its offspring which inherited the variation, be it ever so slight, would also have a better chance. Yearly more are bred than can survive; the smallest grain in the balance, in the long run, must tell on which death shall fall, and which shall survive. Let this work of selection on the one hand, and death on the other, go on for a thousand generations, who will pretend to affirm that it would have no effect, when we remember what, in a few years, Bakewell effected in cattle, and Western in sheep, by this identical principle of selection?'

1.2.2 Wallace's paper to the Linnean Society

Wallace's views were remarkably similar: 'The life of wild animals is a struggle for existence. The full exertion of all their faculties and all their energies is required to preserve their own existence and provide for that of their infant offspring . . . Even the least prolific of animals would increase rapidly if unchecked, whereas it is evident that the animal population of the globe must be stationary, or perhaps, through the influence of man, decreasing. So long as a country remains physically unchanged, the numbers of its animal population cannot materially increase'.

'Most or perhaps all the variations from the typical form of a species must have some definite effect, however slight, on the habits or capacities of the individuals. Even a change of colour might, by rendering them more or less distinguishable, affect their safety; a greater or less development of hair might modify their habits . . . Now, let some alteration of physical conditions occur in the district – a long period of drought, a destruction of vegetation by locusts, the irruption of some new carnivorous animal seeking 'pastures new' – any change in fact tending to render existence more difficult to the species in question, and tasking its utmost powers to avoid complete extermination; it is evident that, of all the individuals composing the species, those forming the least numerous and most feebly organised variety would suffer first, and, were the pressure severe, must soon become extinct . . . The superior variety would alone remain, and on a return to favourable circumstances would rapidly increase in numbers and occupy the place of the extinct species and variety . . . This new, improved, and populous race might itself, in course of time, give rise to new varieties, exhibiting several diverging modifications of form, any of which, tending to increase the facilities for preserving existence, must, by the same general law, in their turn become predominant. Here, then, we have *progression and continued divergence* deduced from the general laws which regulate the existence of animals in a state of nature, and from the undisputed fact that varieties do frequently occur . . .'

'The hypothesis of Lamarck – that progressive changes in species have been

produced by the attempts of animals to increase the development of their own
organs, and thus modify their structure and habits – has been repeatedly and
easily refuted by all writers on the subject of varieties and species, and it seems to
have been considered that when this was done the whole question has been
finally settled; but the view here developed renders such an hypothesis quite
unnecessary, by showing that similar results must be produced by the action of
principles constantly at work in nature . . . (For example) the giraffe did not
acquire its long neck by desiring to reach the foliage of the more lofty shrubs,
and constantly stretching its neck for the purpose, but because any varieties
which occurred among its antetypes with a longer neck than usual at once
secured a fresh range of pasture over the same ground as their shorter-necked
companions, and on the first scarcity of food were thereby enabled to outlive
them.'

'We believe we have now shown that there is a tendency in nature to the
continued progression of certain classes of *varieties* further and further from the
original type – a progression to which there appears no reason to assign any
definite limits . . . This progression, by minute steps, in various directions, but
always checked and balanced by the necessary conditions, subject to which
alone existence can be preserved, may, it is believed, be followed out so as to
agree with all the phenomena presented by organised beings, their extinction
and succession in past ages, and all the extraordinary modifications of form,
instinct, and habits which they exhibit.'

1.2.3   The essence of evolutionary mechanism

Both these papers clearly contain the three facts and two conclusions which
are commonly taken as the simple summary of Darwinism evolution: the
potential of all species to increase greatly in numbers, coupled with an
approximate constancy of numbers, implies that there is a *struggle for existence*;
and when variation is added to this, it is clear that *natural selection* must operate.

It was the ease with which these propositions could be understood that helped
the *fact* of evolution to be generally accepted. In fact Darwin devoted more than
half of the *Origin* to different lines of evidence that evolution has occurred: he
has two chapters on the fossil record, two on geographical distributions, and one
each on morphological likenesses (including comparative embryology, the
interpretation of vestigial organs, and the meaning of classification), behaviour,
and domestication. He devoted later books specifically to the origin of man and
sexual selection, domestication, and adaptations in plants for pollination,
insect-eating, and climbing. All these were parts of the book Darwin had
originally intended to write before being forced into print by Wallace.

1.3   First series of objections: Darwin's own answers

In Chapters 6 and 7 of the *Origin* Darwin deals with 'difficulties' and 'mis-
cellaneous objections' to his theory. His main points concern the nature of
species and questions about the efficacy of selection. In a later chapter, he
discusses the imperfections of the fossil record. Darwin knew that the

maintenance of variation was a key weakness in his theory. The causes of variation are repeatedly referred to in the book and in later editions of the *Origin*, he tended to accept that some Lamarckian explanation might be necessary (*i.e.* that the heredity of an individual might be affected by an environmental modification of its phenotype. No claim of Lamarckian inheritance by Kammerer, Lysenko, Steele, and many others over the years, has ever been substantiated). The problem was not resolved until the physical basis of heredity was discovered following the embryological conclusions of Weissman (1883) and the re-discovery of Mendel's work in 1900. These are dealt with in sections 1.4 and 1.5. The position in palaeontology is reviewed in sections 6.5 and 7.2.

The issues faced by Darwin in the *Origin* are still raised today, and it is therefore relevant to begin with them.

1.3.1 Species transitions

If one species may evolve into another, why are not forms linking two species found? R. A. Fisher has argued that the reason Darwin had to discuss this was the hangover from Lamarckian speculation that existed when he was writing. Lamarck had suggested that evolutionary change arises from the use or disuse of organs and traits, so that transmutation arises from a varying response in an existing group. It would follow from this that there are no firm limits to any species, nor is a 'natural' classification realistically possible since the evolving unit is the individual.

Darwin rejected this idea of species from his personal study of species in nature, and assumed a definition close to the modern one, that a species is an effectively isolated population or group of populations. He recognized that:

(i) Closely-related forms are likely to compete for the same resources, leading to the less-favoured one(s) becoming extinct. This has been shown repeatedly by experiment (perhaps most exhaustively in the flour-beetle *Tribolium*), with the important qualification that different varieties (or species) can survive together only where a heterogeneous environment allows different varieties to occupy different niches. However the point at issue here is that one form will normally become extinct if a selectively advantageous form competes with it.

(ii) In a large area, different species replace each other geographically. In most cases these species seemed to have evolved in isolation and then expanded their ranges to come into contact. Geological changes (separating and recombining tracts of land) play a part here, but at the level of present species the most important factor has been the Pleistocene, when previously widely-distributed forms were isolated in a number of warmer or wetter refuges for relatively long periods, and changed sufficiently to remain distinct when the climate improved and they were able to re-occupy their former territory. Well-worked examples of this process are the Palaearctic ring of *Larus* gull species, and the *Heliconius* butterflies (and their mimics) in the Amazon basin. It is interesting that Darwin recognized the dynamic influence of historical events in forming new species, in contrast to the implicit assumption of both his contemporaries and many recent biologists that environments are stable and largely homogeneous.

(iii) Transitional forms will almost certainly be less common than either the ancestral or descendent form, and hence liable to be overlooked or to become extinct. As collecting and the description of variation has progressed, it has been recognized that some species previously collected only from widely separated areas may be in fact be connected by intermediate forms. We now recognize *clines* of change in particular traits, and also that a species may be polytypic, *i.e.* contain several geographically distinguishable forms which interbreed to a limited extent. Linnaeus confused matters by giving the name 'variety' indiscriminately to geographical races, domesticated races, non-genetic variants, and inherited 'sports'. The idea of a polytypic species radically altered the Linnean concept of a species: the older idea is of a species characterized by a gap separating it from other groups, while a polytypic species is defined by actual or potential genetic continuity between allopatric populations. This does not, however, affect Darwin's point that forms in the process of change are likely to be uncommon. As we shall see, the discovery of active speciation (for example, *Drosophila* flies in Hawaii and cichlid fishes in Lake Victoria: section 6.2.2); the rarity of transitional forms; and the study of hybrid zones (section 5.2) is throwing a great deal of light on the nature and integrity of species.

1.3.2   The effectiveness of natural selection

The most persistent criticism of Darwinism has always been that natural selection is merely a negative instrument removing inefficiency, but incapable of producing novelty or the seemingly perfect adaptation of such features as the eye of a mammal or bird, or the pattern of a butterfly's wing:

1) *Cases of special difficulty* Darwin recognized three situations:

(*a*) An organ (such as the wing of a bat) may be so specialized for its functions as to bear little resemblance to the prototype (the forelimb of an insectivore) from which it must be presumed to have arisen. The difficulty is envisaging a series of organisms with organs of intermediate grades connecting these widely-separated extremes.

(*b*) An organ of extreme perfection (such as the eye in the higher vertebrates) may show such perfect and detailed adaptation that by comparison with the obstacles which the design of such an apparatus would present to human ingenuity, the mind is staggered by the effort of conceiving it as the product of so undirected a process as trial and error.

(*c*) Some organs of seemingly trifling importance (such as the 'fly-whisk' tail of the giraffe) are yet so clearly adapted to the function they perform that they cannot be regarded as accidental. In these cases it may be asked how such a trifling function can ever have been a matter of life and death to the organism, and so have determined its survival in the struggle for existence.

The first of these classes of objection applies to all evolution, whilst the second and third are difficulties more of imagination than of reason. It is impossible to deal with them in detail; R. A. Fisher has commented that 'the cogency and wealth of illustration with which Darwin was able to deal with these cases was, perhaps, the largest factor in persuading biologists of the truth of his views'. Here we can only note that:

(*i*) *Function as well as structure evolves*. For example, there are organisms which have no image-forming eyes, but light-sensitive cells. Any inherited variants which allowed detection of the direction of light, its size, movement, etc., could be of potential advantage, and subject to natural selection. The eye as we know it would be built up by the accumulation of many small steps, each of which could be adaptive. The result is efficiency in a particular environment, not perfection: one could envisage the human eye being 'improved' by functioning better in poor light or under water, or failing to deteriorate with age; these attributes have never been 'necessary' for human survival.

(*ii*) *The usefulness (or 'adaptive value') of a character can be tested by experiment*. Victorian biologists wasted a vast amount of effort speculating about the function or value of particular organs; there is a similar tendency in the 1980s to pontificate about the significance of particular behaviour patterns (section 7.1). Apparently trivial traits may be shown to be highly important: it has only been recently shown that flies may seriously disturb a tropical herbivore, and an efficient 'fly-whisk' may add notably to fitness. Conversely other traits (even the horns of the Irish elk) may be incidental results of selection for other traits associated with growth.

The persistence of these criticisms has come from ignorance, often resulting from lack of research, rather than defects in the underlying ideas.

2) *The origin of novelty*

How can natural selection which functions to filter out deleterious variants lead to completely new developments? Is not natural selection limited to modifying existing adaptations? The answer is no, as consideration of three facts reveals (see also section 6.3):

(*a*) *All* traits are subject to variation.

(*b*) In evolution, novelty is introduced by a change in the environment: animals and plants invaded land because of an available habitat, not because it seemed a good idea. Some characters will be *pre-adapted* to the new environment.

(*c*) Even a very small selective advantage can lead to genetical change (Chapter 4).

In the *Origin*, Darwin quoted Agassiz's work on echinoderms, showing how modification of spines may lead to the development of an apparently new and important trait, tube feet. Many similar examples are known. One of the major biological recognitions of the 1960s was the enormous amount of variation present in virtually every population (Chapter 3), which means that a species can respond rapidly and precisely to new environmental stresses: bacteria can digest oil, aphids detoxify artificial poisons, and plants grow when introduced to the Antarctic Continent. There is no reason for disbelieving that any reasonable novelties cannot occur in evolutionary time. Indeed even apparent *dis*-advantage (such as sterile castes of insects) can evolve in appropriate conditions (section 5.4); natural selection is a mechanism for producing *a priori* improbable contingencies (see below). (This fact, incidentally answers claims that there has not been enough time for evolution to have taken place since the earth became habitable. The commonly used analogy that a monkey randomly typing will produce the works of Shakespeare, but only if he has astronomical time is irrelevant, since selection can rapidly and ruthlessly change the frequency of apparently random variation.)

3) *The strength of natural selection*
   The power of natural selection to produce adaptation could only be illustrated by anecdote until quantitative methods of estimating fitness differences, rates and conditions of gene frequency change, and similar parameters became available. These were developed by R. A. Fisher, J. B. S. Haldane and Sewall Wright during the 1920s, and demonstrated the ability of selection differentials of less than 1% to bring about evolutionary change. The force of their theoretical arguments has been greatly strengthened with the discovery that selection pressures in nature commonly reach 10% or 20% (Table 4) (see also section 4.1).
   Without condoning speculation by protagonists and opponents alike about the possible course of evolution, it can nevertheless be affirmed that answers given by Darwin to his critics in successive editions of the *Origin* have been repeatedly proved right by subsequent research.

1.4   Second series of objections: biometricians and Mendelians

   Darwin had no doubt that inherited variations were extremely common. He had been convinced of this by his observation of varieties in nature, but more especially by his contacts with practical animal breeders and horticulturists. On the other hand, he knew nothing about the mechanism of inheritance. He accepted the conclusions of the botanists Kölreuter (1733–1806) and Gärtner (1771–1850) who crossed vast numbers of plant varieties. Their common finding was that the characteristics of both parents blended in the offspring, which tended to be intermediate between the parents. This meant that if a new variant arose, it would have only half its expression in its offspring, one quarter in the grand-children (because it would almost certainly have to cross with the non-variant form), and so on. New variation would have to arise at a high rate if it was going to persist long enough to be operated on by selection.
   In *The Variation of Plants and Animals under Domestication* (1868), Darwin put forward his 'provisional hypothesis of pangenesis' in an attempt to account for this. He suggested that each part of an organism produces 'free and minute atoms of their contents, that is gemmules' which pass to the reproductive organs and are thence passed to the next generation. 'Direct and indirect' influences of the 'conditions of life' (as suggested by Lamarck) could in this way become part of the hereditary constitution of the organism; as he wrestled with the problem, Darwin found himself more and more adopting Lamarckian ideas.
   Meanwhile Darwin's cousin, Francis Galton, had been carrying out his own investigations into heredity. In 1869, he published *Hereditary Genius*, presenting the pedigrees of numerous distinguished families, and argued from these that talent must be inherited. This was followed in 1889 by *Natural Inheritance* which summarized data on size, disease, temperament, etc. into a general 'law of

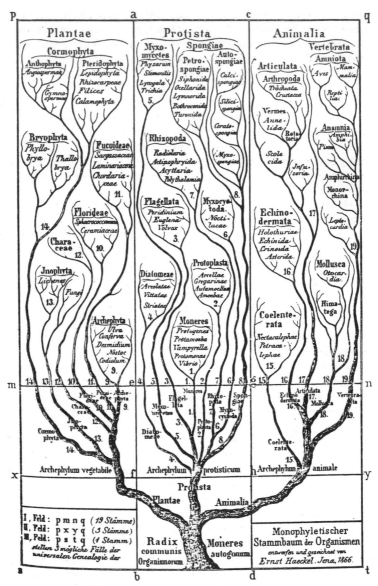

Fig. 1–1 The first phylogenetic tree, drawn up by Ernst Haeckel in 1866.

ancestral heredity', based on the share of inherited constitution an individual transmits to his descendants ($\frac{1}{2}$ to his children, $\frac{1}{4}$ to grandchildren, $\frac{1}{8}$ to great-grandchildren). This was a simple quantitative corollary of the currently-accepted ideas of blending inheritance; but it enabled Galton to develop statistical ideas of correlation and regression, which have remained central to biometrical analysis and quantitative inheritance ever since.

Galton's ideas were seized on by a mathematician, Karl Pearson, and by W. R. Weldon, who was Professor of Zoology at University College London and then Oxford. Weldon realized the importance of Galton's work for evolution, since the frequency of deviations from the type could be measured and associations measured. He believed that 'the questions raised by the Darwinian hypothesis are purely statistical, and the statistical method is the only one at present obvious by which that hypothesis can be experimentally checked', and co-operated with Pearson in seeking methods to test and extend Galton's law. They persuaded the Royal Society of London to establish in 1894 an Evolution Committee 'for the purpose of conducting statistical enquiries into the variability of organisms', and this committee sponsored work on the death-rates of crabs in Plymouth Sound, and measurements of herring and of ox-eyed daisies.

In 1897 the Evolution Committee was joined by William Bateson, a Cambridge zoologist and first director of the John Innes Institute (1910–1926). Bateson was convinced of the importance of discontinuity in evolution, and dubious about the early work of the Committee. He criticized Weldon for measuring crabs at different stages of moult. He rapidly brought about a change in the approach of the Committee, and in 1900 Pearson and Weldon resigned.

Meanwhile the search for a better understanding of the genetical process led in 1910 to the rediscovery of Gregor Mendel's work originally published in Brno, Czechoslovakia in 1866 by the botanist de Vries in Holland, von Tschermak in Austria, and Correns in Germany. Mendel's main conclusions (the regular segregation and independent assortment of inherited traits) were soon confirmed in animals by Bateson and Punnett in Cambridge and Cuénot in Paris.

The application of Mendelian ideas had two important consequences for evolutionary ideas: they indicated that inherited factors did not blend, but persisted unchanged through the generations, thus removing the chronic problem of variation loss; and they appeared to show that inherited variation was discontinuous. In reaction the biometricians (i.e. Pearson and Weldon) regarded Mendelism as a threat. The six years following the rediscovery of Mendelism (until the death of Weldon in 1906) witnessed increasingly bitter confrontations between the Mendelians and biometricians over 'homotyposis' (Pearson's attempt to weld together heredity and differentiation), mutation theory, the meaning of genetical dominance, etc.

The physical basis of heredity was not generally accepted until the work of T. H. Morgan and his colleagues on sex determination, linkage, and mutation in *Drosophila melanogaster* linked breeding results with cytological knowledge (summarized in *The Mechanism of Mendelian Heredity*, published in 1915). However, Mendel's 'laws' (Mendel himself did not state any laws, but his conclusions are most easily summarized in this way) were everywhere being confirmed, and the Mendelians increasingly gained the upper hand. The biometric-Mendelian controversy degenerated into personal conflict, and ended when Weldon died of pneumonia at the age of 46. but the split between the supporters of continuous and discontinuous evolution continued to grow, and was only resolved in the 1930s (section 1.5).

1.5   Third series of objections: palaeontologists and geneticists

In the early 1900s, the importance of natural selection in evolutionary change was believed to be of little importance; the emphasis was increasingly laid on the origin of variation rather than its maintenance. Huxley and Bateson believed that continuous variations were too small to generate significant selection pressures; Galton believed that the selection of continuous variations soon reached a limit because of the counteracting effect of regression. A Danish botanist, Wilhelm Johannsen, showed (1903) that selection for weight within pure (i.e. self-fertilized) lines of beans had no effect, and concluded from this that continuous variation was not inherited and therefore unimportant in evolution. (The significance of the fact that different lines produced beans of different mean weight was not realized at the time, although it is an aspect of Johannsen's work which is almost always quoted nowadays.)

The mutation theory of Hugo de Vries (one of Mendel's re-discoverers) was particularly influential at this time. De Vries worked with the evening primrose, *Oenothera lamarckiana* and observed frequent mutations in his stocks. He argued that evolution depended on such, and that species originated by 'jumps' or saltations rather than the accumulation of small differences as suggested by Darwin (see section 7.2). We now know that de Vries's *Oenothera* mutations were mostly due to chromosomal rearrangements, together with the segregation of recessive traits, and are not mutations in the modern sense. Notwithstanding, a generation of biologists grew up convinced that evolution was 'driven' by mutation, with natural selection taking a minor role.

Meanwhile palaeontologists were building up an increasingly confident picture of evolutionary changes in fossil strata. It seemed clear that much change was continuous and progressive; evolutionary 'jumps' did not exist when the record was continuous over long periods. The mutations being studied by laboratory geneticists appeared to have nothing in common with real evolution. Darwin's own emphasis on gradual evolution was continued by the palaeontologists, *via* the biometricians.

During the 1920s the gap between palaeontologists and geneticists widened. As knowledge of mutations *sensu stricto* increased, it seemed that they almost invariably produced deleterious traits which were inherited as recessives, whereas adaptively useful traits were virtually always dominant. It is no wonder that this period led to a widespread disenchantment with classical Darwinism, and the propounding of a variety of other theories of evolutionary mechanisms: Berg's *Nomogenesis*, Willis's *Age and Area*, Smuts's *Holism*, Driesch's entelechy, and others. None of these was satisfactory, and all depended upon an almost mystical inner urge (or *élan vital*) to progress. It is unfortunate that three standard and still-read histories of biology (by Nordenskiöld, Rádl, and Singer) were written during this time, and the idea that evolutionary theory is an illogical mess has been perpetuated.

1.5.1   Fisher's theory of the evolution of dominance

At the height of the Mendelian-biometrician controversy, a British mathematician, Udny Yule, showed that if the geneticists' assumption that dominance was always complete was dropped, and if the effect of environment on variation was taken into account, the conflict between Mendel's theory and Galton's law of ancestral heredity disappeared. The importance of the environment became increasingly clear from plant breeding experiments by E. M. East and D. F. Jones in New England and others, and from the demonstration by embryologists (E. S. Goodrich in *Living Organisms* (1924), followed by Julian Huxley, Gavin de Beer, etc.) that the primary effect of an inherited factor (a gene) may be considerably modified in expression during development, *i.e.* that it can be confusing not to distinguish between gene and character (or trait).

The question of dominance was taken up by R. A. Fisher. Using Pearson's measurements on man, Fisher dissected the factors involved in characteristics affected by many genes, examining the statistical consequences of genic interaction, assortative mating, multiple alleles, and linkage, upon the correlation between relatives. By comparing the correlations between sibs with those between parents, he was able to distinguish the contributions of dominance and environment to the total variance. This led Fisher to believe that dominance was a result of interaction between genes rather than an intrinsic property of the gene itself, and he speculated on the effect of this in evolution.

Fisher started from the proposition that there is no intrinsic reason to expect a mutation occurring for the first time in the history of a species to be either dominant or recessive; it will most probably be intermediate, with an effect somewhere between its expression in double dose (*i.e.* when homozygous) and the normal (wild-type) condition. However:

1) Mutations will have occurred repeatedly at virtually every locus in past times. The mutations we find nowadays are recurrences of events that have happened thousands of times in the past.

2) When a mutation occurs, it will almost always be present in the heterozygous condition; if an allele has become relatively common in a population, then other forces than mutation must be influencing its frequency (see section 2.1.2).

3) If a newly arisen allele has a beneficial effect on its carrier, combinations of it with other alleles that increase its effect will have a higher fitness than any which decrease it. This will be continually happening so that the genetic architecture of the species will become modified to the extent that any new occurrences of that advantageous allele will always produce the maximum effect in its carrier, i.e. in the heterozygous condition. In other words, beneficial characters will be selected for dominance – and will spread by natural selection to replace the previous expression of the trait. Conversely alleles which are deleterious in the heterozygote are only likely to be transmitted in combinations where their effect is least, i.e. there will be selection for a small heterozygous effect, which means in the direction of recessivity.

Fisher put forward these ideas in 1928 from first principles. They were received sceptically because of the difficulty in believing that selection pressures would be strong enough to allow genes which did nothing else but modify dominance to spread. In pre-1950 days primary selection coefficients were thought of as around 0.1 to 1.0% and second-order effects would be much less than that. J. B. S. Haldane suggested that dominance was more likely to arise from alleles with a margin of safety becoming the normal allele, since they could exercise an undiminished action when heterozygous, i.e. mutant alleles would be recessive and deleterious.

1.5.2   The experimental modification of dominance

Fisher himself was the first to demonstrate the genetical modification of dominance, by crossing domestic poultry to wild jungle fowl for five generations. This changed the inheritance of certain characters so that a degree of heterozygous manifestation occurred where complete dominance previously prevailed. Fisher reasoned that the dominance of the traits he studied had been attained during domestication as a result of selection for the more striking heterozygotes.

A fuller demonstration of the influence of modifying genes on dominance was obtained by E. B. Ford by breeding from the greatest and least expressions of a variable yellow variant (*lutea*) of the currant moth (*Abraxas grossulariata*). Although the differences between *lutea* and *non-lutea* can be regarded as caused by a single allelic difference, after only three generations Ford produced heterozygotes virtually indistinguishable from the *lutea* homozygote in selection for the yellowest individuals, and ones like the typical homozygote in the white selection line. In other words he had changed the heterozygote from a position of no (or intermediate) dominance to one of complete dominance or recessivity respectively. Ford then crossed his modified heterozygotes back to unselected stock, and by the second generation (when the selected modifiers would have a chance of segregating independently) the original variable heterozygotes reappeared. He thus showed that it was the response of the organism rather than the gene itself that had changed. H. B. D. Kettlewell later showed that the 'switch' between the typical and melanic forms of the peppered moth was likewise made up of many modifiers, which had presumably been favoured by selection. Indeed the oldest known melanics caught in the 1850s (which were presumably heterozygotes) had white spots on their wings, showing that dominance was not complete at the time the melanic began spreading (section 5.1).

Fisher's theory has been proved right in a number of similar experiments in both plants and animals, and, although it may not apply for every allele at every locus, it had considerable historical significance in bridging the gulf between palaeontologists and geneticists. In effect the theory provided the genetical basis for the understanding between disciplines which was needed before the neo-Darwinian synthesis could take place.

1.6   The neo-Darwinian synthesis

As late as 1932 T. H. Morgan was asserting that 'natural selection does not play the role of a creative principle in evolution', but ten years later all but a very few biologists were agreed on an evolutionary theory based firmly on Darwin's own ideas knitted with subsequent developments in genetics. This coming together was described by JULIAN HUXLEY as the *Modern Synthesis* in a book of that name, published in 1942. The synthesis can first be seen in three English books, R. A. FISHER's *Genetical Theory of Natural Selection* (1930), E. B. FORD's *Mendelism and Evolution* (1931), and J. B. S. HALDANE's *Causes of Evolution* (1932); it was consolidated in three works from America, DOBZHANSKY's *Genetics and the Origin of Species* (1937), MAYR's *Systematics and the Origin of Species* (1942), and SIMPSON's *Tempo and Mode in Evolution* (1944). As Mayr has written, it did not come about by one side being proved right and the others wrong; but from 'an exchange of the most viable components of the previously competing research traditions'. The dissensions and difficulties of the 1920s and 1930s can in retrospect be seen as a result of increasing specialization by biologists. Darwin himself collected as many facts as he could from the whole of biology, but by the end of the nineteenth century each of the disciplines within biology had grown so much that the glue binding the original synthesis had begun to come apart. The significant movers of the neo-Darwinian synthesis were scientists like R. A. Fisher and E. B. Ford who transferred ideas from one discipline to another; iconoclasts like Cyril Darlington and G. G. Simpson who broke out from their own descriptive disciplines; and individuals like J. S. Huxley and H. J. Muller who moved between research groups. As we shall see, the later 'problems' about evolution in the 1960s and 1980s came largely as a result of ignorance (Chapters 3, 4 and 7).

2 Genetic Forces

2.1 Genetic stability: the Hardy–Weinberg theorem

Evolution is sometimes defined as gene frequency change. This is an oversimplification: evolution may also involve changes in the number of genes or chromosomes (as in polyploidy and gene duplication), or the arrangement of the genetic material (inversions, translocations, movable genetic factors, etc., producing position effect, protected gene sequences, etc.). Notwithstanding, an evolutionary change can be represented at the simplest level as a gene change, and to understand evolutionary forces it is necessary to define how such change frequency comes about.

The starting point is the existence of inherited variation. Take the human blood groups: there are 13 or more blood group systems, determined by at least 20 loci with over 100 alleles. In *families*, it is possible to show that these are simply inherited; in *populations* we are concerned with the frequencies of the groups. Different races have different frequencies.

For example, in 300 native-born Shetlanders typed for the MN system, 104 reacted to anti-M only (which meant they were carrying only the M antigen, and were thus genetically MM), 51 to anti-N only (NN people), and 145 to both antibodies (MN people). In this sample we can count the frequency of M alleles, since MM people carry two each and the heterozygotes (MN) one. There were $[(2 \times 104) + 145]$ M alleles out of 600 alleles in the sample (since every person of the 300 has two alleles).

The *frequency* of M is $[(2 \times 104) + 145]/600 = 0.588$. Similarly the *frequency* of N is $[(2 \times 51) + 145]/600 = 0.412$. Since the frequency of all alleles at the locus must add up to 1.00 (or 100%), we could more easily calculate the frequency of N as $(1.00 - 0.588)$. Conventionally the frequency of alleles in a sample is represented as p, q, etc. In our example, $p = 0.588$, $q = 0.412$, $p + q = 1.00$.

What happens to these frequencies when mating takes place? For simplicity we start with two assumptions:

1) That there is random mating, i.e. that any individual has an equal chance of mating with any other individual in the population, or, put slightly differently, that the chance of fusion between any two gametes is proportional only to their relative frequencies.

2) That the population is large in a statistical sense. In practice this means that it contains more than about a hundred breeding individuals.

Consider the case of a locus with two alleles A, a, such that the frequency of A is p, and of a is q (or $1 - p$). We construct a three-stage argument:

1) The three parental genotypes (AA, Aa, aa) can produce only two sorts of gametes (A- or a-bearing), and the frequency of each gamete will be p and q respectively.

2) The gametes will combine to form zygotes, and we can calculate the frequency of the three classes (AA, Aa, aa).

3) Knowing the genotype frequencies in the children, we can calculate the gene (or allele) frequency in them.

We begin with a 2×2 table (or Punnett square) where the gametes and their frequencies produced by the two parents form the two sides, and the four gametic combinations (which are the children's genotypes) form the body of the table. The frequencies of the filial genotypes can be calculated by the simple product of the gametic frequencies if the two above assumptions hold:

FATHER

| | | A | a |
		p	q
	A p	AA p^2	Aa pq
MOTHER			
	a q	aA qp	aa q^2

In the filial generation, there will be $p^2 AA$, $2pqAa$, and $q^2 aa$. As this includes all zygotes in the sample: $p^2 + 2pq + q^2 = 1$

One point worth making at this stage, is that we can calculate gene frequencies for a recessive character even where we are unable to 'count' alleles as in our Shetlanders. If we can identify only two phenotypes (dominant homozygotes plus heterozygotes, and recessive homozygotes), the frequency of q is simply the square root of the recessive homozygotes in the sample. Thus in our Shetlanders, the frequency of MM was $104/300 = 0.347$; the frequency of M can be calculated as $\sqrt{0.347} = 0.588$. Our previous estimate is theoretically more accurate, because it was not based on assumptions about population size or random mating. However we are forced to use this method of gene frequency estimation if we cannot identify heterozygotes with confidence.

To return to the general case: we know the genotype frequencies in the filial generation, and we can calculate p and q from them by counting. Thus

$$p \text{ (in the children)} = p^2 + \frac{1}{2}(2pq)$$
$$= p(p + q)$$
$$= p \text{ (since } p + q = 1)$$

This leaves us with an apparently trivial conclusion that allele frequencies remain constant from generation to generation. Furthermore, genotype frequencies are determined in a random mating population entirely by allele frequencies, so these will also remain constant.

This result is in fact extremely important. To restate it: *in a large random-mating population, gene and genotype frequencies remain constant in the absence of migration, mutation and selection.* This statement is usually dignified by being called the *Hardy–Weinberg* theorem (or law, or principle). It is really a particular application of the binomial theorem, but most biologists seem to prefer a derivation like the above. The value of understanding the Hardy–Weinberg theorem properly is that it enables the effect of the five agencies which disturb gene frequencies to be invesigated (section 2.2).

2.1.1   The origin of the Hardy–Weinberg theorem

R. C. Punnett (afterwards Professor of Genetics at Cambridge University) has told of lecturing on genetics in 1908 at the Royal Society of Medicine in London: 'I was asked why it was that, if brown eyes were dominant to blue, the population was not becoming increasingly brown-eyed: yet there was no reason for supposing such to be the case. I could only answer that the heterozygous browns also contributed their quota of blues and that somehow this must lead to equilibrium. On my return to Cambridge I at once sought out G. H. Hardy with whom I was then very friendly, for we had acted as joint Secretaries to the Committee for the retention of Greek in the Previous Examination and we used to play cricket together. Knowing that Hardy had not the slightest interest in genetics I put my question to him as a mathematical problem. He replied that it was quite simple and soon handed me the now well-known formula. Naturally pleased at getting so neat and prompt an answer I promised him that it should be known as 'Hardy's Law' – a promise fulfilled in my next edition of *Mendelism*. Whether the battle of Waterloo was won on the playing fields of Eton is still a matter for conjecture; certain it is, however, that 'Hardy's Law' owed its genesis to a mutual interest in cricket.'

In the same year, a German physician by the name of Weinberg published the same result, and we have had the Hardy-Weinberg theorem ever since.

2.1.2   Heterozygous frequency for a rare allele

An important insight into population structure from the Hardy–Weinberg theorem is that the great majority of rare alleles will be carried in the heterozygous condition. It will be remembered that this was one of the points in Fisher's theory of the evolution of dominance (section 1.5.1). For example, albinism in man is a recessively-inherited condition; its frequency in Britain is about 1 in 20 000.

$$\text{Hence } p^2 = 0.000\,05$$
$$p = \sqrt{0.000\,05}$$
$$= 0.0071$$

The expected frequency of heterozygotes is
$$2pq = 2 \times 0.0071 \times (1 - 0.0071)$$
$$= 0.0141 \text{ or } 1 \text{ in } 71$$

In other words, heterozygotes are 280 times commoner than homozygotes.

The chance of two heterozygotes marrying will be $(0.0141)^2$ or one chance in 5000 (one in four of their children will be albinos, which brings us back to the population frequency of the condition, 1 in 20 000). There is one important qualification to this risk of two heterozygotes marrying: if two people share a common ancestor in their recent history, they will have a much greater chance than average of both being heterozygotes for the same allele. Although the frequency of cousin marriages in Britain is less than 1%, 8% of albinos are the children of cousins.

At once we are introduced to two significant properties of natural populations: firstly, uncommon alleles will exist almost entirely in the heterozygous state, and this means their effect in the population as heterozygotes will be proportionately much greater than as homozygotes. Secondly, alleles are unlikely to be distributed at random through a species because individuals are more likely to mate with a neighbour than a far distant member of the species, and the neighbour is proportionately more likely to share a common ancestor (Fig. 2–1).

2.2 Forces changing gene frequencies

The five agencies which may affect a Hardy–Weinberg stability are: sampling error in small populations, mutation, natural selection, migration between groups, and mating choice.

2.2.1 Small populations (genetic drift)

The gametes that transmit genes from one generation to another carry a sample of the genes in the parent generation. If the sample is not large, gene frequencies are liable to change between the generations. This can be understood by thinking about the human sex ratio. There are very nearly the same numbers of boys and of girls born in the world. Yet there are many single sex families – two, three, four, or even more children of the same sex. However, for every four boy family, somewhere there is a four girl family; overall the sex ratio evens out. The larger the group we are considering, the more likely it is that we shall find equal numbers of boys and girls. The important question in the present context is what happens in small populations, because however many individuals there may be in a species, in a local mating group there may only be comparatively few breeding pairs. For example, the common wren (*Troglodytes troglodytes*) is widespread and fairly common throughout northern Europe and western Asia, but it has a number of small isolated populations. There are about 100 pairs on St Kilda and under 50 on Fair Isle.

Gene frequencies in small populations like these may behave in the same way as sex ratio in humans. When Mendel crossed the first generation progeny of pure-breeding pigmented and non-pigmented pea plants, he obtained 705 pigmented and 224 non-pigmented ones. This was not quite a 3:1 ratio, although gametes carrying the pigmented and non-pigmented alleles would have been produced in *exactly* equal number. If he had bred a smaller number, he might have found even greater 'sampling error'.

The theory behind this sampling error is simple. In any generation the frequency of an allele will be distributed about the mean value (p) with a variance $p(1-p)/2N$, where N is the number of organisms breeding in the population (and $2N$ the number of gametes from which they were formed). Consider populations with 2, 20, and 200 individuals respectively, each with a frequency of p of 50%. The standard errors of p will be respectively 25%, 7.9%, and 2.5%;

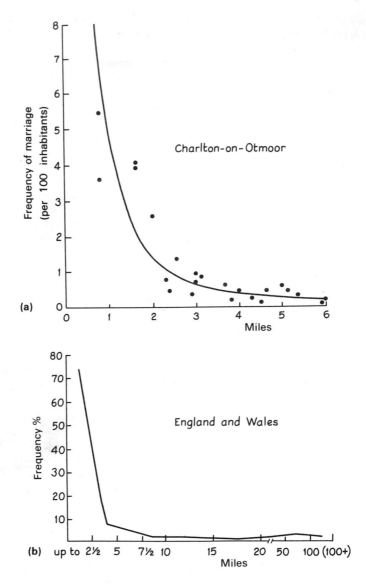

Fig. 2–1 Most matings take place between close neighbours. These figures show the average distance apart of marriage partners in: **(a)** the Oxfordshire village of Charlton-on-Otmoor in the mid-nineteenth century, and **(b)** the whole of England and W̧ales between 1930 and 1960 (omitting large towns). Similar curves have been obtained from the foraging of bees; the flight of *Drosophila* over several days; cross-pollination by bees; and the wind dispersal of spores or pollen. (From BERRY, 1977.)

as the population size increases, the value of p will show less and less fluctuation. In the smallest population, p may change to 25% or 75% by chance; whereas in the largest population p will only vary between 47.5% and 52.5%.

The fluctuation from generation to generation is called *genetic drift* (or simply *drift*), or sometimes the Sewall Wright effect after the American geneticist who derived its theory in great detail. It has three consequences:

1) If a population remains small, frequencies may change erratically in a 'bagatelle' manner, and the rarer allele at any locus will inevitably be lost in due course. The probability of an allele being lost in any generation is $1/2N$. This means that genetical variability will decrease.

2) As alleles are lost, there will be fewer heterozygotes; and as inbreeding increases, the proportion of homozygotes will rise.

3) A series of small populations formed from a single large one will inevitably diverge, even if they were initially identical.

The trouble with drift is that it is the easiest way of explaining differences between allele frequencies in two populations, and it has been repeatedly claimed for cases of divergence. Darwin himself suggested that 'I am inclined to suspect that in some at least polymorphic genera (such as *Rubus*, *Rosa* and *Hieracium* amongst plants, several genera of insects, and brachiopod shells), we see variations which are of no service or disservice to the 'species' and . . . (are) not affected by natural selection, . . . left either as a fluctuating element or (one which) would ultimately become fixed.' A classical example of speculation that has been proved wrong (see section 6.2.2) is that speciation in land snails on high volcanic islands like Tahiti has resulted from random divergence (i.e. drift) of groups of snails separated from each other in the steep valleys running radially from the central volcanic peaks. Although drift has been shown in laboratory conditions, no good example has ever been proved in nature.

Intermittent drift   However a form of random genetic change which has been called 'intermittent drift' has certainly been significant in evolution (Fig. 2-2). This is when the numbers of a population are drastically reduced by a catastrophe or in the process of founding a new population (well seen in the formation of island races). The descendent race will almost inevitably differ from the ancestral one in the frequencies of alleles at many loci, and show a stochastic difference in a single generation. The 'founder effect' is the most drastic way of changing gene frequencies that there is.

2.2.2 Mutation

Mutation is a rare event, occurring at the rate of about one in 10 000 loci per generation, although it may be increased several-fold by chemical or physical influences (the most well-known of which is ionizing radiation). A mutational change alters allele frequency, but the amount and speed of change is trivial when compared to other allele-changing forces.

2.2.3 Natural selection

The effect of natural selection is measured by reproductive success or failure. Clearly death before the completion of breeding may be selective, but *any* allele

Frequencies of allelomorphs at one locus in a population

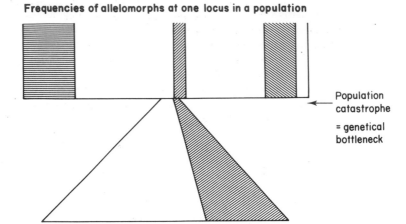

Fig. 2–2 Intermittent drift: a drastic method by which allele frequencies change and variation is reduced.

which affects the number of children produced will be represented in the next generation by more or fewer copies, and thus be subject to selection. Selection can be defined formally as *the differential change in relative frequency of genotypes due to differences in the ability of their phenotypes to obtain representation in the next generation.* It may affect only one extreme phenotype (*directed selection*), as in the spread of melanism in moths living in industrial areas; both extremes (*stabilizing selection*), reducing variation in favour of the mean value of a character; or the mean only (*disruptive selection*), with the result that the extremes increase at the expense of the average. Natural selection is by far the strongest agency changing gene frequencies, and we shall return to it repeatedly, especially in chapters four and five.

2.2.4 Migration between groups

Two populations are unlikely to be genetically identical, but the more interchange of individuals between them, the more they will come to be alike. Put another way, a population may gain an allele either by mutation in one of its own members or by immigration (or *gene flow*) from a population which contains it. However the most important effect of migration is probably to retard or prevent local adaptation.

2.2.5 Mating choice

One of the assumptions made in deriving the simple Hardy-Weinberg situation was the random union of gametes – that the formation of zygotes is uninfluenced by any characteristic of the parents. The most obvious exception to this is sex, because members of one sex can only mate with members of the other (or others: in the self-incompatibility system of flowering plants there may be up to 100 'sexes' available). However, mating between either like or unlike phenotypes may be a means of maintaining variation in a population. Penrose has suggested that human intelligence is affected by such assortative mating, since intelligent people tend to marry intelligent people, and unintelligent people similarly marry unintelligent people. An allele in the scarlet tiger moth (*Panaxia dominula*) affecting wing pattern remains segregating in some colonies despite reducing fecundity and longevity, for the simple reason that females prefer mates with a different phenotype to their own (Fig. 2–3).

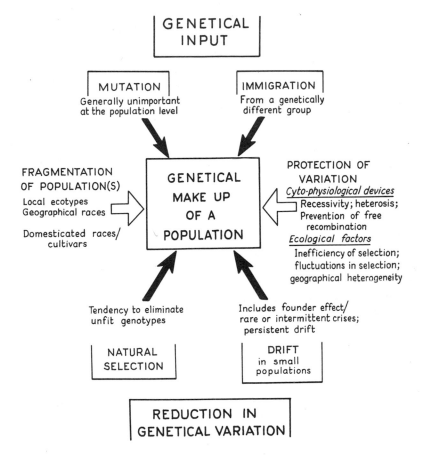

Fig. 2–3   Summary of the factors contributing to the genetic makeup of a population. (From BERRY, 1977.)

2.3   Genetic equilibria

There are two common genetic equilibria in populations.

2.3.1   *Mutation* versus *selection*

If a mutation is deleterious to its carrier, it will be transmitted to the generation at a lower rate than the normal allele, and will in due course be eliminated from the population, unless it recurs through fresh mutation. For example, human achondroplasics have normal fertility and intelligence, but on average only one-fifth the number of children of non-achondroplasics. The frequency of this dominantly-inherited condition is constant, meaning that 20% of achondroplasics are born to an affected parent; the other 80% are the result of fresh mutation.

A similar balance may be established between immigrant gene flow and selection. Mutation and immigration are the only ways of increasing the genetic variation to a population; drift and selection reduce genetic variation.

2.3.2   *Heterozygous advantage and polymorphism*

If a heterozygote has a higher fitness than the two homozygotes (or the mean value for a continuous trait when compared to extreme values), a stable genetic polymorphism will be set up. The simplest way of envisaging this is where both homozygotes at a locus with two alleles are lethal, so only heterozygotes will survive and mate. The allele frequencies of both alleles will be 50%. More usually the homozygotes will have a reduced but not zero fitness. In this case, the frequency of the alleles will be determined by the relative fitness of the two homozygotes.

Simple heterozygous advantage at a single locus is probably uncommon, but the same effect is produced if one homozygote is selected against at one stage of the life cycle or in one niche, and the other at another stage or niche.

Heterozygous advantage (or *over-dominance*, or *heterosis*: the terms are effectively synonymous) maintains genetical variability, and for single loci may produce different genotypes at relatively high frequencies. If these are easily distinguishable they are called *morphs*. E. B. Ford has defined genetic polymorphism as *the occurence together in the same locality of two or more discontinuous forms of a species in such proportions that the rarest of them cannot be maintained by recurrent mutation* (note that the definition excludes geographical variation). The importance of the recognition of genetic polymorphism is that it directs attention to polymorphisms as worth investigating for the detection of evolutionary forces.

2.4   Genetic load and the cost of selection

During the 1950s theoretical arguments seemed to show that there were limits to the amount of variation that could be carried by a population or species.

2.4.1   *Genetic load*

H. J. Muller was one of T. H. Morgan's original *Drosophila* colleagues (section 1.4). In 1927 he devised the C1B technique for measuring the genetic effects of radiation. In the years that followed he observed many times the effect of radiation in inducing deleterious mutations in his fly cultures, and found that even the heterozygotes of 'recessive' alleles had a reduced fitness of $2\frac{1}{2}\%$ on average. Since the heterozygotes are very much commoner than the homozygotes, virtually the entire effect of these mutations will be exercised in the heterozygous condition, and it is easy to show that they will persist for an average of 40 generations before being eliminated by selection.

Muller was worried about the effect of an increased mutation rate in man. He argued that if the number of genes in man is 5000 and if the mutation rate per gene per generation is 1 in 50 000, then the average number of detrimental genes in the heterozygous state per individual is: 2 (because we have two chromosome sets) × no. of loci × mutation rate × persistence = 8.

This crude estimate of detrimental genes has been supported by studies on human populations in which frequencies of death and inherited disease in the children of cousins has been compared with that in the children of unrelated parents. If it is assumed that all the excess abnormality in cousin marriages is due to rare alleles becoming homozygous, it is possible to calculate the average number of deleterious recessives we carry. Estimates in different populations are in the range three to eight per person. Returning to Muller's argument, he maintained that the mean fitness in man is reduced by $(8 \times 2\frac{1}{2}\% =) 20\%$ as compared to the maximum fitness of a person carrying no mutants. He called this the *genetic load*. It means that the average number of surviving children is only 80% of those that might have been born, and hence at least 2.5 children per couple must be conceived if the population size is to remain constant.

Muller believed this genetic load was effectively the result of recurrent mutation. It follows that if the mutation rate doubles, only 60% of the children conceived would survive. This would mean that each couple would conceive 2.5 children, but only produce 1.5 living adults for the next generation. The population would not replace itself, and would rapidly become extinct.

2.4.2 The cost of selection

J. B. S. Haldane called attention in 1957 to the *cost* of natural selection. This occurs whenever one allele spreads in a population. The allele that is being replaced will result in a lower fitness for its carriers than the allele that is spreading. If an allele spreads to fixation (i.e. if the former allele is completely eliminated in a population whose size remains constant), the number of individuals that die through natural selection will be equal to nearly two-thirds of those alive in any one generation. The advantage produced by the new allele does not matter: an allele that spreads rapidly (i.e. that has a high selection advantage) will be responsible for a higher proportion of selective deaths in fewer generations than an allele spreading more slowly. Clearly, however, a population can only tolerate a limited number of substitutions going on at the same time, or too many individuals will die to maintain a viable population. Haldane calculated that one allele can be substituted every 30 generations on average, and supported this conclusion by deductions from the rate of change in fossil lineages.

Haldane's cost represents a *substitutional* load in Muller's terminology; Muller's own emphasis on the effects of mutation ought to be expressed as a *mutational* load. There is also a *segregational load*, caused by the relatively lower fitness of some genotypes in balanced polymorphisms. All these parts of the total load contribute to the idea that there is an upper limit to levels of genetic variation.

2.5 Continuous variation

The early biometricians believed that continuous variation had a different genetic basis from discontinuous. Anyone with an elementary knowledge of heredity now knows this not to be so, and the same considerations of genetic change and stability apply to multigenic as to monogenic traits. The main difference lies in the techniques of analysis that have been developed (largely by plant and animal breeders); at the moment continuous variation has to be approached through the mean, variance and heritability of particular traits, although the time is approaching when the contributions of individual genes to continuous traits will be assessable (see section 4.4).

3 Neutralism and Selection: a Further Synthesis

In the early 1960s, evolutionary theory seemed to have been thoroughly worked out by the theoreticians, and little remained except details. However in 1966, two papers were published which seemed to challenge the whole neo-Darwinian edifice. We can regard this as the fourth set of objections to evolution.

3.1 Electrophoresis and heterozygosity

Electrophoresis involves running an electric current through a protein-containing extract in a gel (usually made of starch or a polyacrylamide ester). The net charge on the protein molecule will determine its direction and speed of movement in the electric field; if two proteins differ only in their net charge, they will separate on the gel. One-third of amino-acid substitutions in a protein (i.e. gene mutations) will change the net charge on the molecule, and it is therefore possible to identify in a protein extract the homozygotes and heterozygote for approximately a third of all segregating genes affecting protein structure. If a number of proteins are studied from a random population sample, it is thus possible to measure objectively the amount of genetic variation in that population. Using this technique, Lewontin and Hubby in Chicago working with *Drosophila pseudoobscura*, and Harris in London working on man independently showed that both species had so much variation that they ought to be extinct: three out of 10 loci were polymorphic in man, and a mean of 9.9% were heterozygous; in *Drosophila* seven out of 18 were polymorphic, and each fly was on average heterozygous at 11.5% of its loci. This result has been extended to several hundred species, and has been confirmed virtually every time. A few species have a low level of variation, but the normal range of mean heterozygosity ranges from approximately 3% in large animals to 25% in small ones, with plants generally around 15–20% (Table 1).

The absolute values of variation obtained in this way have been criticized on grounds such as that atypical and soluble proteins are usually studied, but this does not alter the main conclusion that genetical variation is far more common than previously assumed. The theoretical basis of neo-Darwinism seemed to be destroyed.

3.2 Neutralism

The simplest way out of the variation dilemma was to assume that the protein variants detected had no effect on the properties of their protein; in other words that they were selectively neutral. The main positive argument for this conclusion was the alleged accuracy of the 'protein clock'.

The 'protein clock' is based on the chemical assumption that the chance of

Table 1 Genetic variation as measured by electrophoresis.

		Average number of loci scored per species	Mean % of loci	
			polymorphic per population	heterozygous per individual
Insects	Drosophila	24	52.9	15.0
	Other insects	18	53.1	15.1
Molluscs	Land snails	18	43.7	15.0
	Marine snails	17	17.5	8.3
	INVERTEBRATES		39.7	14.6
Fish		21	30.6	7.8
Amphibians		22	33.6	8.2
Reptiles		21	23.1	4.7
Birds		19	14.5	4.2
Mammals	Rodents	26	20.2	5.4
	Large Mammals	40	23.3	3.7
	VERTEBRATES		17.3	5.0
	PLANTS		46.4	17.0

substitution of an amino acid at any place in a protein chain is random, and therefore the number of differences in protein composition between two species can be regarded as a measure of the time since the species shared a common ancestor, allowing the construction of a genealogical tree showing the degrees of relationship. When this is done, there at first appeared to be remarkable uniformity in the rates of substitution along completely different evolutionary lines. For example, the ancestors of mammals and fish diverged 350 to 400 million years ago. When the differences in the haemoglobin molecules of carp and man are added up, it can be calculated that the average rate of change since they diverged from each other has been 8.9×10^{-10} per amino acid site per year (i.e. about one substitution every 10 million years). Comparisons among mammals from different orders which radiated about 80 million years ago produce a substitution rate of 8.8×10^{-10}. The β chain in the globin molecule probably originated as a duplication of the α chain. Both carp and man have β chains, so the duplication must have occurred at least 400 million years ago. When the α and β chains of human haemoglobin are compared, they show a rate of change of 8.9×10^{-10}. Thus the speed of divergence of α and β chains seems to be about the same as independent estimates of the rate of change over 350 million years and over 80 million years.

However different proteins show rates of change differing by nearly 100-fold. Even worse, the rate of change of the haemoglobin α chain of the early vertebrates was faster than in mammals; and the rate slowed markedly in the line giving rise to the apes. Biochemists have speculated about 'accelerated' or 'retarded' rates at different evolutionary stages, and introduced concepts of 'functional constraints' or 'dispensability' of some parts of protein molecules. These ideas have been valuable in probing the functional morphology of some well-studied proteins (notably the cytochromes, globins, and immunoglobulins), but destroy any idea of an accurate protein clock (Table 2). From the

Table 2   Rates of protein evolution. The 'unit evolutionary period' is the average time in years $\times 10^6$ for a 1% difference in amino acid sequence to arise between two lineages.

Protein	Unit evolutionary period
Histones: H4	400
H2	60
H1	8
Collagen	36
Albumin	3
Casein	1.4
Fibrinopeptide	1.4
Glutamate dehydrogenase	55
Lactate dehydrogenase	20
Triosephosphate isomerase	19
Carbonic anhydrase	4
Electron carriers:	
Cytochrome c	15
Cytochrome b	11
Ferrodoxin	6
Hormones:	
Glucagon	43
Corticotrophin	24
Insulin	14
Prolactin	5
Growth hormone	4
Myoglobin	6
Haemoglobin α	3.7
Haemoglobin β	3.3

evolutionary point of view, the simplest assumption is that some substitutions are advantageous and some deleterious, and have exactly the same properties as any other allelic changes.

A rather similar argument to the 'protein clock' one has been based on the connection between speciation and 'genetic distance' (i.e. a measure of the amount of difference between two groups in terms of the frequencies of alleles at a number of (usually) electrophoretically detected loci. Obviously the differences between local races are less than between species or genera (Table 3). The genetic distance argument is biologically better than the protein clock one, because it takes into account the circumstances of the populations concerned rather than the mere passage of time. Nevertheless, it is wrong to argue that speciation takes place merely because of some threshold level of genetic separation (see section 6.2).

However, one of the underlying assumptions of the protein clock argument was disproved when it was shown that substitutions in proteins are not random but involve chemically similar amino acids more often than would be expected by chance. This result is entirely consonant with natural selection: more distinct amino acids are less likely to be successfully substituted, for their

Table 3   Genetic similarity at early stages of evolutionary divergence.

	Local populations	Sub- or semi-species	Species	Genera
INVERTEBRATES:				
Drosophila willistoni	0.97	0.79	0.47	
Drosophila obscura	0.99	0.82		
Drosophila repleta	1.00	0.88	0.78	
VERTEBRATES:				
Lepomis (Sunfish)	0.98	0.84	0.54	
Cyprinidae (Minnows)	0.99			0.59
Taricha (Salamanders)	0.95	0.84	0.63	0.31
Sceloporus (Lizards)	0.89	0.79		
Mus (House mice)	0.95	0.77		
Dipodomys (Kangaroo rats)	0.97		0.61	0.16
Peromyscus (Deer mice)	0.95		0.65	

chemical properties may interfere with the functioning of the molecule, and hence the fitness of the individuals concerned.

However the final collapse of the extreme neutralist position has come from the recognition that many (and perhaps most) variants change the properties of the enzymes in which they are substituted, and are thus potentially subject to selection. Studies have been carried out on biochemical properties *in vitro*, on correlation of variant distribution with environmental variables (particularly temperature), and on changes of allele frequency in time and space (and in both field and laboratory). Some of these results are described in Chapters 4 and 5. The firm conclusion is that allele frequencies of all sorts are liable to fine adjustment by natural selection.

3.3   Why be variable?

It is all very well to show that the observed variation in populations is non-neutral, but the question then becomes the more interesting one as to how variation is maintained.

1) Both Haldane and Muller effectively thought of every gene acting independently on its carrier. This is patently not true. The difference in fitness between an individual heterozygous at (say) 20% of its loci and one heterozygous at 10% is unlikely to be great, and extreme multiple heterozygotes will be so rare that they will not contribute significantly to a normal population. Selection is likely to pick out individuals with relatively high frequencies of heterozygous loci, and eliminate the tail of predominantly homozygous individuals. For a given level of mortality, this will produce typically stabilizing selection. In other words, the entire phenotype will be selected, not independent aspects of it.

2) Selection pressures change in time and space. This means that fitness is not a constant, but depends on both the physical and biological environment (see section 4.4). Looked at from this point of view, the 'segregational load' would be better called the 'environmental load' since it depends on selection exerted against one or both homozygotes, and the strength of this selection is directly related to environmental pressures. In the classical example of the sickle cell haemoglobin polymorphism, the advantage of the heterozygote disappeared in negro slaves taken to North America where there was no malaria. The frequency of the sickle cell allele is now declining there entirely through (directed) selection against the anaemic homozygotes.

In formal terms, selection pressures may be density or frequency dependent or independent. For example, a low density of a prey species attracts neither predators nor intra-specific pressures, and a struggle for survival only intensifies when numbers increase. It is also possible to distinguish between 'hard' and 'soft' selection. Under hard selection, an individual has to meet certain requirements to survive. For example, a marine organism is unable to cope with the osmotic problems of living in fresh-water; most plants are poisoned by the soil concentrations of heavy metals on mine spoil heaps; unselected mosquito populations cannot tolerate a high concentration of insecticide; and so on. In contrast 'soft' selection begins to operate only when the normal carrying capacity of an environment is filled.

When we examine the situations where genetically caused deaths occur, we are faced with the enigma that a genetically invariable population has no 'load', but may be faced with extinction if the environment deteriorates without new variants appearing. This is true of the classical case of the spread of the melanic form of the peppered moth (section 4.1), which Haldane used as his example of the cost of selection. The species only survived in polluted areas because of the occurrence of melanic mutants.

In situations where an environment changes and reduces the chance of survival of a species living in it, a 'load' is placed upon the species which has nothing to do with genetics. A gene replacement process offers an escape from this external load. If the replacement is rapid enough, the gradual decrease in environmentally-caused deaths will enable the population to survive while another population without an alternative allele may succumb to the changed environment.

Although different genetical causes of death all contribute to 'genetic load' and can all be described by the same algebra, they differ considerably in their contribution to the species. Mutational load – the original Mullerian load – is bad; the elimination of mutants is a maintenance job to keep the population in working condition. Substitutional load is definitely favourable to the population, although it may coincide with a time of increased mortality if it occurs in response to an environmental crisis. Segregational load is a paradox. It can result in clearly identifiable genetical deaths, yet its overall effect on the species is to increase its tolerance to environmental variation. Perhaps it is offensive to our sense of fairness: it would be more democratic to have everyone sickly than to have a few die outright.

Much of the difficulty of the load concept is due to the fact that it wrongly combines the mean and variance of a population's viability through a false assumption that increase in the number of segregating genes must be associated with decreased mean fitness. It would be an advantage to keep the fitness and variance of a population distinct: the mean fitness involves biological and ecological factors of considerable complexity, while the statistical variance of fitness can be dealt with unambiguously if the question of population 'success' is omitted.

3) Population composition cannot be separated from population history. In many ways the genetic composition of a population is a record of its history through past environments and vicissitudes. Some alleles will respond to current environmental pressures, but others will merely reflect past events. For example, it has been suggested that human blood group frequencies were determined by the major epidemic diseases of the past; the wing pattern in an isolated population of the marsh fritillary butterfly (*Melitaea aurinia*) became extremely variable at a time when the population size increased, and settled to a new pattern when numbers again decreased. An allele may only be adaptive once in a life-time, or even once in 100 generations, but will long persist as an indicator of a past adaptive episode.

4) There may be genes with only a small primary effect on the phenotype, but which exercise an important role as modifiers of major gene effects. This would mean that the net selection pressures exercised on any such loci which are segregating would be much lower than those indicated by studies on major polymorphic genes. Indeed the simple ideas of genetic polymorphism which have been the mainstay of ecological genetics are far too naive when we consider gene interactions in development and behaviour.

3.4   The fourth synthesis

Looked at rather harshly, much of the support for the neutral mutation point of view came from the elegance of the mathematical arguments for it, rather than their necessary correctness. This does not mean that the neutralist criticisms of neo-Darwinism were wrong. Probably both extreme neutralists and extreme selectionists were over-persuaded. Some mutations will have no effect on fitness for much of their life; the crucial point is that neutralism of such alleles is unlikely to persist in the long term. However the neutralist–selectionist episode served two very useful functions:

1) It broke the theoretical stranglehold on evolutionary biology which had developed during the 1950s and threatened to turn biology into a branch of applied mathematics, and

2) The converse of this is that biologists have been recalled to biological phenomena in the widest sense (biochemical, immunological, behavioural, and physiological, as well as ecological and morphological), and seen the need for resisting the temptation to seek all their answers in the molecular laboratory. The significance of this in the context of this book is that disputes about evolution have repeatedly turned out to be repeats of old battles, fought in ignorance of relevant facts and arguments. We shall find exactly the same situation when we consider the fifth series of objections to evolution in Chapter 7.

4 Natural Selection

The significance of two sets of data have not yet been fully assimilated by those who criticize neo-Darwinian orthodoxy. The first is the discovery discussed in Chapter 3 of large amounts of variation in populations, which finally destroyed the old idea of a population or species 'type' homozygous for wild type alleles at all but a very small proportion of mutant or polymorphic loci. Evolutionary adjustment is not dependent on or driven by new mutations, but has an enormous resource of potential phenotypic variability. This was assumed by classical geneticists whose conclusions demanded a large supply of gene 'modifiers' to explain observed responses to selection, but had no objective basis until electrophoresis was applied to population samples of a large number of proteins in the late 1960s and 1970s.

The other relevant data are results from the genetical study of natural populations showing that strong intensities of selection are relatively common (section 1.3.2). One of the ironies of the Darwinian saga is that virtually no attempts were made to investigate the power of natural selection; palaeontologists and biologists alike concentrated on phylogenetic speculation. Even today much ink is wasted by both evolutionists and their critics as to the likelihood of past evolutionary events, rather than spending time on the experimental study of evolution. It is only comparatively recently (and due in some measure to what an American has called 'the fascination with birds and gardens, butterflies and snails which was characteristic of the pre-War upper middle class from which so many British scientists came') that ignorance about actual genetical pressures in populations has been overcome.

4.1 Strengths of selection

Elementary textbooks nowadays contain many examples of selection: industrial melanism in cryptically-coloured moths, sickle-cell haemoglobin in malarial areas, inherited resistance to pesticides, tolerance of grasses to high concentrations of heavy metals on mine spoil heaps, and many others (see also Study in Biology, no. 69 *Genetics and Adaptation* by E. B. Ford). These examples are sometimes criticized as involving only details of intra-specific variation, and being irrelevant to the origin of new species. We shall return to this point in Chapter 6. From the present point of view, the important fact to note is that selection coefficients are often very strong. To put these in perspective: one of the most rapid evolutionary changes on record is the doubling of human skull capacity (= brain size) between *Australopithecus* and modern man. However this represents a mean selection pressure of only 0.04% per generation, and is thus easily attainable. It can be compared with the stabilizing selection for birth weight in human babies (i.e. increased mortality of light and heavy babies when compared with the mean), which was 2.7% in the 1930s, falling to about 1.3% in the 1970s as perinatal survival improved (i.e. as the environment changed, in this case as the result of improved medical care.).

The history of knowledge of industrial melanism is an instructive episode in the understanding of the effectiveness of natural selection. It has been most studied in the peppered moth (*Biston betularia*). Black individuals were first caught in Manchester in 1848 as the grime of the industrial revolution began to affect the lichen-covered surfaces on which the typical form was disguised. Melanics were reported from Cheshire in 1860, Yorkshire in 1861, Staffordshire in 1878, Norfolk and Suffolk between 1892 and 1895, and finally London in 1897. Haldane calculated the increase in Manchester from 1% of the melanic in 1848 to 95% in 1898 represented a 33% disadvantage of the typical form when compared to the melanic. The spread of the melanic form seemed to be connected with its appearance, but there were doubts as to how this worked: in the 1930s E. B. Ford showed that melanics in many species had a higher viability in the laboratory when starved, and argued that this difference allowed them to increase in frequency when their cryptic disadvantage was lessened by the spread of pollution.

Kettlewell compared the survival of melanic and typical individuals in polluted and in unpolluted areas by mark-release-recapture experiments and found a much higher survival of the cryptic form, whichever this was. However it was not until he filmed insectivorous birds taking the more conspicuous individuals from tree-trunks that bird predation was accepted as a major influence in regulating gene frequencies. This demonstration has been followed by a considerable volume of behavioural work on the ability of predators to distinguish between small differences in their prey, and no-one now doubts the efficacy of this form of natural selection.

Another object lesson in the power of natural selection has been the work of Dobzhansky on chromosomal inversions in *Drosophila pseudoobscura*. The common laboratory species *D. melanogaster* only rarely possesses inversions, and when they were discovered in *D. pseudoobscura* there appeared to be no pattern in their occurrence or distribution, and since all the flies looked alike, it was assumed to be random, i.e. a consequence of genetic drift. However, repeated sampling of natural populations showed that there was a regular seasonal change in frequencies at any one locality, and also a regular geographical pattern (e.g. with increased altitude in the Sierra Nevada of California). These could only be explained by strong selection pressures. But the clinching evidence for natural selection came from the finding that inversion frequencies in laboratory cultures were susceptible to particular environmental conditions, such as temperature, food, and competition (e.g. crowding); equilibrium frequencies are reached in a particular environment, and these change in a predictable way if the environment is altered (Table 4).

4.2   Selection varies in time and space

Natural selection may change, stay, or split phenotype frequencies. In more conventional language, selection may be directed, stabilizing, or disruptive (see

Table 4   Intensities of selection in natural populations.

Selection for:	% strength of selection
A Directed selection (i.e. extreme phenotype affected)	
Heavy metal tolerance of grasses on mine spoil heaps	46–65
Non-banded *Cepaea nemoralis* in woodlands	19
Melanic (*Carbonaria*) form of *Biston betularia* in various regions of Great Britain	5–35
Spotted form in over-wintered Leopard frog (*Rana pipiens*) (*v* unspotted)	23–38
Unbanded water snakes (*Natrix sipedon*) (*v* heavily banded)	77
B Stabilizing selection (i.e. intermediate phenotype favoured)	
Coiling in snails (*Clausilia laminata*)	8
Size and hatchability in duck eggs	10
Birth-weight and survival in human babies	2–7
Inversion heterokaryosis in *Drosophila pseudoobscura*	Up to 50
Tooth variability (i.e. *de*-stabilizing selection) in *Mus musculus*	21–26
Shell variability in *Nucella lapillus*	0–91
Colour morphs in *Sphaeroma rugicauda*	50 +

section 2.2.3). Directed and stabilizing selection are the more familiar modes: the spread of melanic peppered moths or rats resistant to Warfarin poison are examples of directed selection; the maintenance of egg mimicry in cuckoos or biting efficiency in mosquitoes are examples of stabilizing selection (as is the *maintenance* of melanic peppered moth and warfarin-resistant rat frequencies). Generally speaking, stabilizing selection will be commoner but less intense than directed selection, since it is concerned with preserving the *status quo* while directed forces will be related to rates of change in the environment.

4.2.1   Disruptive selection

From the point of view of evolution, disruptive selection produces two different phenotypes from a single one and therefore leads to divergence, the beginning of the origin of new species. Historically little attention was paid to it, although the idea was familiar to palaeontologists concerned with diverging lines of fossils.

The theory is simple; where there has been considerable argument is the relation between gene flow and the establishment of stable equilibria by random mating. If the population is a single breeding unit, selection against heterozygotes (because that is what disruptive selection is) gives an unstable equilibrium; and any displacement from that unstable point will lead to the eventual loss of one of the alleles from the population. Disruptive selection *per se* is not a sufficient condition for the production or maintenance of a stable equilibrium – or more important in the long run – the development of reproductive isolation between divergent forms.

What is relevant is to decide on the conditions where divergence *can* arise between parts of a population being selected for different phenotypes. Traditionally population divergence was believed to arise only when a single population was physically split into separate subpopulations with reduced gene-flow between them, thus allowing each population to adapt by normal directed selection. Maynard Smith has shown mathematically that there are possibilities of achieving stable equilibria by disruptive selection, but that it is difficult – population sizes in different parts of the population must be separately adjusted, and selection must be strong (of the order of 30%). While neither condition is particularly unlikely, the combination is sufficiently restrictive for differentiation to be commonly based on disruptive selection *plus* other forces. Notwithstanding isolation is not a prerequisite for divergence, and selection between incompletely isolated groups has been shown to produce divergence in the laboratory in a variety of organisms (*Drosophila* spp., *Tribolium, Zea*, and *Brassica*).

Batesian mimicry is a situation in nature in which disruptive selection is important. The best strategy for a mimic species is to develop a number of forms, each resembling a different model. Once the disrupting situation has occurred, frequency and density-dependent selection will take over and increase the resemblance of model and mimic.

4.2.2   Inconstant selection

A classical example of stabilizing selection was in sparrows (*Passer domesticus*) collected cold and exhausted after a severe snow-storm in New England. Out of a total of 136, 72 survived and the rest died. Nine measurements (body weight, total length, wing span, etc.) showed that it was the more variable birds which failed to survive. In other words, the extremes perished, and the average lived to pass on their genes.

The opportunity to study selection in this way is not common, but with a minimum of equipment it is possible to investigate selection in action by comparing the properties (behaviour, physiological tolerance, habitat, fecundity, etc.) of genetic morphs. Any naturalist will be aware of many examples: colour or pattern in many woodlice, cuckoo-spit insects, spiders, ladybirds, land and littoral snails, sticklebacks; heterostyly in primroses, purple loosestrife, campions; flower colour in foxgloves, thistles; leaf coiling and spotting in cuckoo pint (lords and ladies); radiate and non radiate flowers in groundsel; growth characteristics of grasses on different soils; and many others.

The interest in such polymorphic situations is that they show different morphs are adapted to slightly different environments, i.e. they are maintained by selection which has different effects on different members of the population. We have already noted that selection may vary with population density or morph frequency or with occasional environmental fluctuations; the general lesson is that natural selection is an opportunistic agent which may influence *any* genetical variation at any time.

4.3   Favourable mutations

A recurring complaint about Darwinism is that evolution cannot depend only on chance mutations since these are almost always harmful to their possessors.

Part of the answer to this criticism came from Fisher's ideas on dominance modification (section 1.5.1), but it is worth emphasizing that a small but significant proportion of mutations are in fact advantageous, despite being random changes to a functioning organism.

Clear signs that a mutation is non-deleterious occur whenever a new mutant spreads in a population. The best evidence for this comes when an environment changes (often through human activity). For example, the rosy minor moth (*Miana literosa*) was extinct in the Sheffield area for many years until a melanic mutant appeared in the mid-1960s, and the species recolonized the city centre.

However, quantitative data about favourable mutations have largely come from experiments designed to measure the hazards of ionizing radiation. For example, Bruce Wallace monitored the number of recessive lethal alleles in *Drosophila melanogaster* cultures given 2000 rads of γ rays per generation. After 60 generations an equilibrium between induction of new mutation and elimination by selection was reached with about 80% of second chromosomes carrying an allele which would be lethal in the homozygous state. Despite this, the irradiated flies had an average fitness only 2% below that of the controls: their 'genetic load' from mutation was almost entirely compensated for by some mutants which must have produced a high heterozygous fitness (even though they were deleterious when homozygous).

Confirmation of this conclusion comes from other experiments where induced mutants reached and remained at high frequencies in experimental populations long after irradiation or other artificial stimuli are removed. Moreover competitive ability and capacity to respond to selection pressures (for longevity, fecundity, etc.) are significantly greater in irradiated than non-irradiated flies.

No one doubts that most mutations are harmful, but conversely there should be no doubt that advantageous mutations are continually arising and can be favoured by selection. Natural selection is far and away more than a simple sieve to remove genetic damage.

4.4   Evolutionary genetics

There are only four factors which can change the size (N) of a population: births (B), deaths (D), immigrants (I), and emigrants (E).

$$N = B - D + I - E$$

B, D, I and E are all subject to inherited variation; the traditional ecological exercises of determining a life-table or fitting population numbers to simple theoretical models (such as those derived from the logistic equation) disguise a host of processes as different genotypes react differently to environmental or temporal pressures. Ecologists are, of course, aware that the population parameters they measure are genetically determined, but they tend to assume that intra-population variation is negligible, or, alternatively that selection will have maximized fitness. In other words, they burden themselves with a population type as stultifying as the species type of the classical taxonomists. This over-simplification is not tenable in view of the large amounts of inherited variation and strong selection pressures discussed in previous sections. The fitness of an individual or genotype must be regarded as a variable in exactly the same way as, say, birth rate or gene frequency.

Now the fitness of an individual is the outcome of all its developmental and physiological processes; strictly speaking fitness is not a genetic trait but an epigenetic one, i.e. resulting from the interaction of gene-controlled processes (Fig. 4–1). In a few cases the contribution of different genes to fitness has been worked out. For example, two genes have been identified as contributing to resistance to a disease called foul-brood in honey bees: one determines 'larval cell uncapping' and the other 'larval removal'. Wild-type alleles at both loci are

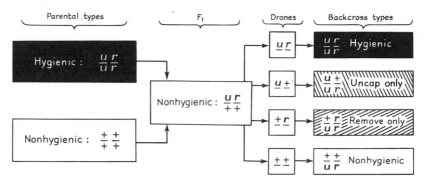

Fig. 4–1 Fitness is an interaction between many genes and the environment. Resistance to foul brood disease in bees is determined by two genes; only the doubly recessive workers (u/u, r/r) will both uncap cells containing dead larvae, and also remove the larvae. (From BERRY, 1977.)

necessary for hygienic behaviour, which protects the colony from the disease. The effect of gene action on behaviour affecting fitness is also well shown by genes involved in tyrosine metabolism: black rabbits are less timid than agouti ones, permitting longer feeding and hence survival in food-limited populations despite visual selection against them; melanic cats are more successful at breeding under dense urban conditions than lighter animals; dark phase Arctic skuas (or jaegers) pair more readily and therefore earlier than ones of pale or intermediate phase; melanic rock doves breed all the year round, and hence increase in frequency when food is permanently available, as in towns; human phenylketonurics (who are mentally defective unless treated) are less pigmented than their normal sibs (Table 5).

And so on: it would be easy to extend greatly the list of inherited fitness variation. From the evolutionary point of view the point to recognize is that fitness variation is extremely common. Ecologists have developed an interest in recent years in what has come to be called 'evolutionary ecology', involving the breeding success of organisms in different niches and how they adjust to change. However, evolutionary ecology is unlikely to advance far unless it is linked to a study of what may be called 'evolutionary genetics'. Criticisms about the effectiveness of selection are inevitable if the action of selection in natural conditions is not properly understood. The popular understanding of selection

Table 5 Genetical influences on population number determinants (after R. J. Berry 1979) in 20th British Ecological Society Symposium, *Population Dynamics*, edited by R. M. Anderson, B. D. Turner and L. R. Taylor. Blackwell, Oxford).

Phenotype	Selective agent
1 Birth	
Aggressiveness in *Mus musculus*	Breeding success
Transferrin polymorphism in several mammals	Infertility
Embryonic mortality from *t*-alleles	Distorted segregation ratios
Nest selection in *Cuculus canorus*	Imprinting on host
Breeding season in melanic *Columba livia*	Steroid control of breeding season
Colour phases in *Stercorarius parasiticus*	Food availability and pair formation
Clutch size in birds	Food collecting ability
Melanism in *Adalia bipunctata*	Rate of heat absorption
2 Death	
Human haemoglobinopathies and other polymorphisms	Epidemic disease, especially malaria
Energy mobilization in *Apodemus sylvaticus*	Food shortage?
Haemoglobin and blood efficiency in *Mus musculus*	Low (winter) temperature
Esterases and chick colour in *Lagopus scoticus*	Population density
Pesticide resistance in rodents, mosquitoes, aphids, pathogenic micro-organisms	Direct human offensive
Melanic moths in industrial areas	Visual predation by birds
Cepaea nemoralis colour and banding morphs	Visual predation; climatic tolerance
Sphaeroma rugicauda colour morphs	Temperature
Spirorbis borealis settling behaviour	Substrate and population density
Heavy metal tolerance in marine invertebrates and grasses	Toxic levels in the environment
Stabilizing selection on morphological traits: *Passer domesticus, Nucella lapillus*, etc.	Climatic stress
3 Immigration and emigration	
Persistence of small mammals in refuge habitats	Differential migration
Population fluctuation in *Microtus* spp. and allozyme frequencies	Aggression and dispersal?
Spread of *Streptopelia decaocto*	Temperature tolerance??
Reduction in flight frequency in *Amathes glareosa*	Life on a small, windy island?
Outbreeding in angiosperms	Heterostyly, self-incompatibility, etc.
Poor survival of metal-tolerant grasses in normal sward	Intolerance of crowding

tends to be restricted to a relatively few fairly crude examples, such as bird predation in industrial melanism or pesticides and resistance pests. The important endeavours for both geneticists and ecologists in the coming years should be to dissect the contribution of different genes to births, deaths, immigration and emigration. This will involve collaboration with biochemists, physiologists and developmental biologists; the answer to the objections of the 1860s, 1900s, 1920s and 1960s all came from syntheses wrought across specialisms. It would be sensible to expect new advances in evolutionary biology to proceed from analogous collaborations.

5 Genetic Architecture

The combination of high variability and strong selection produces rapid and precise adaptation. Such fine genetic tuning can only be appreciated if a population is repeatedly sampled. This is much more difficult than surveying geographical or ecological variation, and has been done much less frequently. Nevertheless there are now many examples showing change in gene frequencies as the environment changes: chromosome inversions in *D. pseudoobscura* changing seasonally have already been mentioned; other examples are bird selection against different colour morphs of the snail *Cepaea nemoralis* in woodland as the green background of spring gives way to the brown of autumn; change in temperature sensitive morphs of the isopod *Sphaeroma rugicauda* and the ladybird *Adalia bipunctata* as winter cold strikes; and even alternation in physiologically different haemoglobin morphs at different times of the year in a mammal (the house mouse). Such precise response makes more believable the perfection of adaptation which has been commented on ever since the days of Darwin (section 1.3.2). However evolution is not simply a juggling with different alleles to produce an identikit fit to different environments; it is concerned with real animals and plants gathered into species with vastly different life-cycles, powers of dispersal, and capacities of change. Before reviewing our knowledge of speciation (Chapter 6) we must consider some of the factors that retard adaptation.

5.1 Genetic plumbing

The distribution of alleles is rarely constant throughout the range of the species. When a change of gene frequency occurs, what Julian Huxley named a *cline* is said to exist. For example the frequency of black peppered moths declines from central Liverpool into rural Wales; the frequency of bridled guillemots increases steadily from southern Britain to Iceland; the incidence of cyanide-containing white clover falls off in Europe from a maximum in the warm south and west; and so on. A cline may arise through:
(*a*) Random genetic drift producing a frequency difference in one part of a population's range.
(*b*) Contact between two genetically distinct populations.
(*c*) Spatially discontinuous changes of environment.
(*d*) Continuous environmental gradients.
It may be stable or transient; a transient cline may result from re-establishing a stable situation, or the spread of an advantageous allele which proceeds like a wave through the population. We can call all these genetic plumbing.

5.2   Hybrid zones

The simplest model of a cline is of alleles spreading like oil on water, with the speed of spread depending on the amount of individual movement before mating. However a uniformly distributed population probably does not occur in real life, and it is better to think of species being divided into a number of *islands* or *stepping-stones* with a finite amount of gene-flow between them. Using the plumbing analogy we would expect the connection between two such populations to behave like a partially clogged pipe, and merely slow down gene-flow. In fact stable hybrid zones are not infrequently formed at the boundary of two races; they are known in baboons, bats, rodents, lizards, frogs, fish, snails, grass-hoppers, butterflies, beetles, stick insects, ticks, and many species of plants. The question to be asked is how two forms can breed together and produce completely viable offspring, but not completely blend with each other.

In extreme cases there is no problem about this. For example when a horse and a donkey cross to produce a mule, the mule is sterile because differences between the chromosome sets from the two parents prevent normal meiosis. In many cases the distinction between the two parents is much less. For example carrion and hooded crows (*Corvus corone corone* and *C.c. cornix*) meet in a narrow hybrid zone across the central Scottish Highlands in which pairing seems

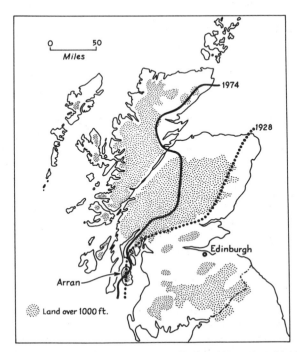

Fig. 5–1   The hybrid zone between carrion (to the south and east) and hooded (to the north and west) Crows in Scotland (solid line); the zone has moved northwards in the Eastern Highlands. (From BERRY, 1977.)

to be at random. The offspring of mixed unions can be recognized by the amount of grey colouring on the back and undersides. The range of the hoodie seems to have contracted somewhat in recent years, but the two crows remain completely distinct over most of their range (Fig. 5–1). The same conservatism occurs in an area through central Denmark where grey- and white-bellied races of house mice meet. The extraordinary thing here is that North American mice are descended from the white-bellied form of Europe, and are more like them than is the European white to the European grey although these two are in breeding contact.

The factors producing conservatism in a population are described as *coadaptive*. We can also speak of them as resulting from *genetic architecture*. This implies the interdependence of different variants in the same way as a human architect is limited by the materials at his disposal as well as the needs of his client.

5.3 Features of genetic architecture

Consider two loci A, B affecting the same or associated characters, with each locus having two alleles (A,a ; B,b), and assume that A, B influence the character(s) in question in one direction, while a, b influence it in the other. Genes affecting human height and sternopleural chaetae in *Drosophila* are examples of such systems (although these examples probably involve at least five to 10 genes). The two loci can give rise to nine genotypes, and if we give A, B 'values' (i.e. effect on the phenotype) of one, and a, b 'values' of 0, then:

	Value		Value		Value
AA BB	4	Aa BB	3	aa BB	2
AA Bb	3	Aa Bb	2	aa Bb	1
AA bb	2	Aa bb	1	aa bb	0

If the two loci are on different chromosomes, the chances of extreme phenotypes will depend only on the frequencies of the segregating alleles; if the loci are linked, there will be an additional restriction provided by the amount of recombination between the two. If we make one further assumption that extreme expressions of the phenotype are disadvantageous to its possessor in a particular environment, important consequences follow.

The character we are concerned with produces the greatest fitness for its carriers when it has a value of two. This can be produced by two homozygous and two heterozygous genotypes:

$$
\begin{array}{cccc}
A\;b & a\;B & A\;B & A\;b \\
\dfrac{+\!\!-\!\!+}{+\!\!-\!\!+} \quad \text{or} & \dfrac{+\!\!-\!\!+}{+\!\!-\!\!+} \quad \text{or} & \dfrac{+\!\!-\!\!+}{+\!\!-\!\!+} \quad \text{or} & \dfrac{+\!\!-\!\!+}{+\!\!-\!\!+} \\
A\;b & a\;B & a\;b & a\;B
\end{array}
$$

The first two (homozygous) genotypes have no inherited variation and might lead to extinction if the environment changes; the two (heterozygous) genotypes both combine present fitness with the possibility of future variation, and will be of greater long-term benefit. However, the second heterozygous genotype (where the loci are 'in repulsion') will only produce extreme (0,4) offspring if there is recombination between the loci; the other double heterozygote (which is 'in coupling') will give extreme offspring every time genetically like gametes fuse. So linkage in repulsion will combine both fitness and flexibility. Any mechanisms (such as chromosomal inversions or translocations) which promote the accumulation of chromosomes heterozygous for balances (for 'high' and 'low' factors) will be favoured.

The main evidence for balanced chromosomes comes from artificial selection experiments: repeatedly it is observed that selection for a particular trait leads to decreasing viability (as the chromosomes become 'unbalanced'), but viability is restored if selection is relaxed for a few generations (giving the chromosomes time to recombine and re-establish the balanced state), after which further selection may be possible and effective – indeed there may be accelerated responses as particularly favourable recombinants are produced.

5.3.1 Linkage disequilibrium and supergenes

Selection for genetic architecture leads to certain combinations of alleles being present more often or more rarely than expected. This is described as *linkage disequilibrium*. For example, colour and banding loci are close together on the same chromosome in the snail *Cepaea nemoralis*, but in most places the majority of snails tend to have either a pink-banded or a yellow-unbanded shell; strong linkage disequilibrium occurs between electrophoretically detected loci in some cereals.

Dobzhansky's work on the distribution and maintenance of inversions in *Drosophila* species was described in section 4.1. Dobzhansky found that two inversions may be strongly heterotic in flies collected from the same area, while flies with the same inversions but from different areas are much less likely to show heterosis when crossed. Although an inversion may have a wide distribution, it acquires extra mutations which will be locally tested and proved for their contribution (or not) to local survival. Just as the roofs of houses in northern climates tend to be steep so that snow slides off them, so the genetical architecture of a species has to accommodate to varying conditions throughout its range.

Supergenes are particular examples of linkage disequilibrium. Complex behaviours or patterns often depend on their wholeness and are inherited as a single unit, which on analysis turns out to be a group of closely-linked loci. Examples of 'super-genes' are the chromosome segments controlling heterostyly in primroses (Fig. 5–2), Batesian mimicry in swallow tail butterflies, and the major histocompatibility complex (HLA) in man.

5.3.2 Dominance modification

Dominance involves genetic architecture, because once it has evolved, the population will possess modifiers to ensure that an allele always manifests as a dominant or recessive trait. The best known examples of this are in moths and grass-hoppers.

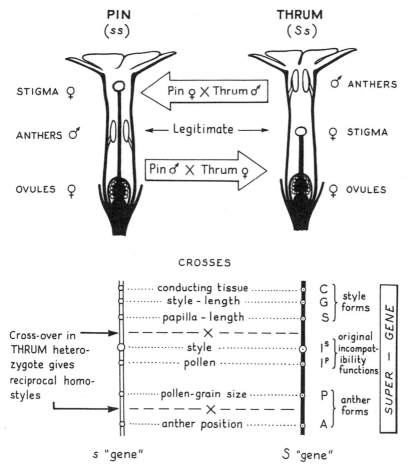

Fig. 5–2 Example of a supergene: heterostyly in the primrose (*Primula vulgaris*) is inherited by a linked complex of at least seven different loci. Recombination within the supergene may lead to homostyly. From BERRY, 1977.)

5.4 Kin selection and altruistic behaviour

A consideration of breeding structure led to the discussion of evolutionary conservatism and genetic architecture; it also tends to the question of behavioural evolution.

From Darwin onwards, attempts have been made to explain the evolution of behaviour beneficial to the group rather than the individual; geneticists have had to emphasize repeatedly that it is individuals and not groups who survive and breed, and that there is no such thing as group selection, i.e. selection for traits beneficial to the group at the expense of the individual. The dilemma is classically expressed in terms of a person prepared to risk himself for others, thereby putting his 'good behaviour' genes at risk, contrasted with a person with 'selfish' genes who does not hazard himself, or his genes.

However it is now accepted that altruistic behaviour can evolve if the helper and the helped are related. Maynard Smith has called this *kin selection*, 'the evolution of characteristics such as self-sacrificing behaviour (like injury-feigning in birds or sterility in social insects) which favour the survival of close relatives of the affected individual'. The next stage has been to define 'inclusive fitness', which adds the investment an individual has in his own genes to those he shares with his living relatives. Kin selection will favour the activities of an organism which benefits its relatives, and act against those which harm them. It is effective because virtually all natural populations are heterogeneous mixes of families, rather than randomly-mixed genotypes. Its consequences can be seen in a wide range of territorial and population-regulating mechanisms in which behaviour apparently acts to benefit the group at the expense of the individual, and where the aggressive bully does not necessarily triumph.

6 Species and Macroevolution

The evidence for infra-specific genetic adjustment (*micro-evolution*) is over-whelming. Populations have to be thought of as being in a state of dynamic genetic flux with their inconstant environments; the old idea of a genetic type manifested in a fixed taxonomic or museum type is clearly invalid. However, there is much less certainty about whether the well-understood processes that produce differences between populations within a species also lead to the formation of new species, and whether 'higher' taxa (genera, families, classes, etc.) are the result of the same processes again. Part of the problem is the practical difficulty of recognizing the limits of a species. Although it is acceptably defined as 'a group of actually or potentially inter-breeding natural populations which is reproductively isolated from other such groups', or 'the largest and most exclusive reproductive community which shares a common gene pool', in practice:

1) Species are more often than not described on the basis of a few dead specimens, and hence depend on morphological criteria.

2) Species exist and vary in both time and space. Confusions arise because of lack of knowledge about the extent of variation within different species, and because of gradations between forms which on some criteria would be regarded as a distinct species. Both these problems apply to interpretations of the fossil record, since such species have to be described almost entirely on morphological grounds. (The study of fossil assemblages or *palaeoecology* is beginning to contribute further information, but is still in its infancy.)

6.1 Species are evolutionary units

Species are keys in the evolutionary process for two reasons:

1) They are natural units because all members of a species share a common gene-pool, thus separating them from other species (even though a very limited amount of gene-flow may take place through the hybrid zone type of situation described in section 5.2. Groups which share gene-pools in this restricted way can be described as *semi-species*; they are an interesting but fairly unimportant exception to Darwin's emphasis on the discreteness of species described in section 1.3.1). The existence of natural units with varying degrees of relationship make possible a natural (or phylogenetic) classification.

2) Being independent units means that their survival, change, and extinction are constrained by their ecological and not their genetical associations. Evolution above the species level is, of course, still dependent on inherited variation, but the emphasis moves to survival of the whole group in its particular environment(s).

6.2   The mechanism of speciation

Species may split to form two species (and thus increase diversity) or change through time so that the descendent species is so different from its ancestor that it is given a new name. Traditionally speciation through splitting has been regarded as almost invariably by geographical isolation from each other of two (or more) parts of a gene-pool (*allopatric speciation*), allowing differences to accumulate in the separated portions to the point where they become intersterile if they meet again. The most important exception to speciation following isolation was formerly believed to be polyploidy, which has been an important speciating mechanism in higher plants. (Differentiation between sexes has prevented polyploidy being important in higher animals.) Older discussions relegated other *sympatric* speciation (i.e. splitting within a single gene pool) to a minor role, and argue about the significance of *parapatric* (or *stasipatric*) speciation (i.e. within a continuous cline, *q.v.* section 5.1).

An obvious difficulty about investigating speciation is that genetic differences between species can rarely be studied directly. Nevertheless a widely-quoted statement in 1974 by the American biologist Richard Lewontin that 'we know virtually nothing about the genetic changes that occur in species formation', is a pessimistic and misleading exaggeration.

6.2.1   Reproductive isolating mechanisms

The essential elements in speciation are mechanisms which prevent successful hybridization. These can be divided into:

1) *Prezygotic* which impede hybridization between members of different populations, and thus prevent the formation of hybrid zygotes.

2) *Postzygotic* which reduce the viability or fertility of hybrids.

Pre- and postzygotic mechanisms all reduce gene exchange between populations, but there is a waste of reproductive effort if a zygote is formed and fails to develop, which does not occur if there is a hindrance to zygote formation. The apparently high fertility of hybrids between species or orchids, cichlid fishes, and ducks suggests that speciation in such forms entails the development of strong pre-mating but only weak post-mating barriers to gene exchange: in orchids, the highly species-specific scent, colour pattern, and form of the flowers results in them being visited by only a very few species of insects, upon which the pollen masses are deposited in precise positions; in cichlids, movement and species-specific colour patterns are both critical for successful mating; among birds, the most extravagant developments of species-specific male coloration, accessory plumes, etc., occur in groups such as birds of paradise and humming-birds that do not form a pair bond.

Artificial selection against hybrids can lead to a rapid increase in mating discrimination: this has been shown in both *Drosophila* and maize. The most probable sequence of events in nature is that a reduced fitness of hybrids through some form of disruptive selection leads to the development of pre-zygotic barriers. Post-zygotic incompatibility will not be reinforced by natural selection since any fertile hybrids will propagate their genes (a possible exception to this is

Table 6   Reproductive isolating mechanisms

EXTRINSIC: may or may not involve genetical factors
  1 *Geographic separation*
  2 Other spatial separation (e.g. certain parasites)

INTRINSIC: has a genetical base
  1 *Asexuality*, etc.
  2 *Premating or prezygotic* (prevent the formation of hybrid zygotes)
    a Ecological or habitat isolation
    b Seasonal or temporal isolation ('mate avoidance' e.g. in flowering or hatching time)
    c Sexual or ethological isolation ('mate rejection')
    d Mechanical isolation (i.e. structure of genitalia or flower parts prevents copulation
        or transfer of pollen)
    e Isolation by different pollinators (related species of flowering plants may attract
        different insects as pollinators)
    f Gametic isolation (in organisms with external fertilization, gametes are not
        attracted to each other; with internal fertilization, the gametes or gametophytes of
        one species may be inviable in other species).
  3 *Postmating or postzygotic*
    a Hybrid inviability (hybrid zygotes have reduced viability or are inviable)
    b Hybrid sterility (F_1 hybrids fail to produce functional gametes)
    c Hybrid breakdown (F_2 or backcross hybrids have reduced viability or fertility)

---

where a zygote is cared for by its parent, since hybrid zygotes can be discriminated against by the parent). In a few organisms, the effects of individual genes on pre- and post-zygotic barriers have been worked out. For example, there is inherited variation in *Drosophila melanogaster* for mating speed, the duration of copulation, and the extent of assortative mating, all of which contribute to prezygotic barriers.

6.2.2   The speed of speciation

A well-worked example where parapatric (if not sympatric) speciation has taken place involves species of the land snail *Partula* on the volcanic Moorea island in French Polynesia. These were originally regarded as the result of random differentiation (through genetic drift) on isolated parts of the island (see section 2.2.1) but it is now clear that an early colonizing group separated into races locally adapted to different habitats, independent of geographical isolation, and that these have diverged to form at least nine different species capable of some gene-flow but in general behaving as distinct species.

The best examples of speciation occur where island populations have significantly changed from their mainland ancestors. Darwin himself was impressed by the diversity of fauna on the Galapagos archipelago. For example, there are thirteen species of ground finches (Geospizinae) which have adapted to different habitats. Presumably this adaptation occurred on different islands, and then specialized finches successively colonized other islands; there are now ten finch species on the main island of Indefatigable. In contrast, there is only one finch species on Cocos, a single high tropical island where there has been no opportunity for allopatric adaptation and divergence. There are many other examples of island species of birds in Australasia, lizards in the West Indies (where the original divergence took place on land now largely submerged, and represented now by groups of islands which share species), and moths in South America (which survived the Pleistocene in wet 'islands' surrounded by desert).

The most elegant reconstructions of animal speciation come from the Hawaiian chain. The Hawaiian islands are the summits of ocean-bed volcanoes extending eastwards from the Emperor Seamounts in the north-west Pacific, marking the southward and then (for the last 40 million years) eastward movement of the Pacific tectonic plate. There are four main peaks remaining above sea-level, the easternmost (Hawaii Island) being about two-thirds of a million years old. Work has been done on the lobeliads, the marine fauna, birds (especially honey-creepers), but most spectacularly on the insects: one-third of the world's 1500 *Drosophila* species are endemic to the Hawaiian chain. The relationships between different groups of these (especially the hundred or so 'picture-wing' species, which have a body 5–7 mm long) have been unravelled using differences between species in the polytene chromosomes. Most of the species can be bred and mated *inter se* in the laboratory, so all the tests of species can be made. Although there are fairly marked morphological differences between picture-wing species, many of them hybridize easily in the laboratory, and produce fertile offspring. However species have characteristic mating behaviours (often involving 'lek' displays by the males) which produce strong pre-mating isolation.

In the picture-wings, all the polytene chromosome bands are recognizable in every species. There is no chromosomal polymorphism, but many species can be distinguished by unique inversions. Using these it is possible to derive a phylogenetic tree, which shows that there have been 39 successful colonizations between islands. Although some allopatric speciation may have occurred, most of the present diversity has arisen sympatrically.

The Hawaiian *Drosophila* situation focuses attention also on the speed of speciation. The maximum age of the endemic species is the geological age of the island where they are found, and clearly many of the species will have a much more recent origin. The recognition that many of the species are established by a group of colonizers moving from one island to another, with a distinct gene pool (distinct from its ancestors through the founder effect, section 2.2.1), coupled with evidence that has accumulated in the past few years on the speed of adjustment to new environments (section 4.1), shows that speciation may be a process measured in tens rather than thousands of generations. Another Hawaiian example which can be dated particularly accurately involves a banana-feeding moth genus (*Hedylepta*): five endemic species have been described, but bananas were only introduced to Hawaii 1000 years ago. Other dateable events associated with speciation are the formation of a lake cut off from Lake Victoria in Uganda 4000 years ago, where the new lake contains five species of cichlid fish, each similar to but distinct from one of the species in the main lake; distinct house mice in the Faroe Islands, although they could not have got there before man arrived about 1000 years ago; 14 endemic species of cyprinid fishes in a Filipino lake formed 10 000 years ago.

Part of J. B. S. Haldane's conclusion on the cost of selection (section 2.4.2), was that if species differ by alleles at 1000 loci, a speciation event would take 300 000 years. We now know that this estimate is far too high. Indeed there is on record a strain of *Drosophila paulistorum* which was fully interfertile with other strains when first collected from the wild, but developed hybrid sterility after being isolated in a separate culture for just a few years. With the exception of polyploidy, species do not arise at a single step, but genetic differences producing reproductive and/or morphological distinctiveness at the species level can develop very quickly indeed in terms of the evolutionary time-scale (Fig. 6–1).

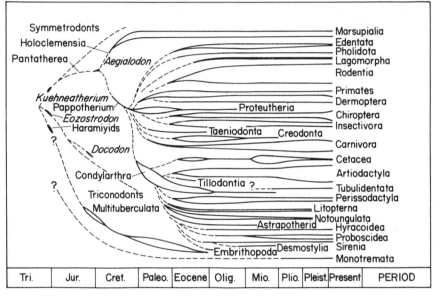

Fig. 6–1 Rates of evolution vary enormously at different times. This figure shows the explosive appearance and diversification of mammalian orders in the late Cretaceous and Palaeocene. (From Ayala and Valentine, 1979, *Evolving*. Benjamin Cummins.)

Linked with the speed with which a species can split to form two, are the existence of species swarms (implicit in the above examples; there are many others). They underline that speciation may occur at very different rates at different times. Palaeontologists recognize this by distinguishing *bradytelic* (slow) and *tachytelic* (fast) evolutionary rates, and by describing changes in the diversity of faunas and floras at different epochs. Evolution is not a remorseless and steady progress towards perfection, but an opportunistic response to new habitats and diversification as fresh possibilities appear. It is intriguing to hark back to Darwin's account of speciation in the *Origin*, and to recall his emphasis on environmental change as the initiative towards speciation (section 1.3.1), and contrast this with the relentless evolutionary advance assumed by palaeontologists in the 1920s (section 1.5).

6.2.3 Speciation in animals; speciation in plants

Animal speciation differs from plant speciation:

1) animals can easily develop reproductive isolation based on behaviour patterns, which will be reinforced by natural selection. In contrast, the possibility for establishing pre-zygotic barriers in plants (particularly wind-pollinated ones) is obviously much less.

2) Because of their great capacity for vegetative growth, sterility is much less of a hindrance to plants than animals. For example, the Canadian pond-weed (*Elodea canadensis*) has spread throughout European water-ways from an introduction in 1840 of a few female plants only. Many plant species (e.g. dandelion, *Taraxacum officinale*) reproduce almost entirely asexually.

3) As already noted, sterile hybrids of plants may acquire fertility by polyploidy more easily than can animals. For this reason, 'reticulate evolution' (i.e. repeated splitting and fusion of lines) is more common in plants than animals.

6.2.4 Genetic differences between species

Older biologists used to speculate about the number of genes differing between species. When electrophoresis gave the possibility of scoring the alleles present at a large number of loci, some workers attempted to devise a quantitative measure of genetic separation to define sub-species, semi-species, species, genera, etc. This approach has failed utterly, because two species may be different at only a few loci directly producing reproductive isolation, or very many which involve sterility incidentally. In the same way, chromosome number and total arm length may vary little in a group, or enormously between two closely-related species. For example, all the living hominoids (chimpanzee, gorilla, orang, and man) have very similar karyotypes, the main difference being that a centric fusion has occurred to produce a metacentric chromosome in man ($2n = 46$) which is absent in the other three ($2n = 48$); in contrast a race of Madagascan lemur (*Lemur macaco collaris*) has at least four different karyotypes in different parts of its range ($2n = 60$, two different forms of 52, and 48); the Indian muntjac (*Muntiacus muntjac*) has $2n = 7$ in the ♂, $2n = 6$ in the ♀, while its close relative, the Chinese muntjac (*M. reevesi*) has $2n = 46$. Gene and chromosome studies lead to the same conclusion as investigations of molecular change (section 3.2), that evolution is not a product of linear and constant adaptation, but proceeds by fits and starts; just as the environment may change in both the short and long term and just as selection is an inconstant force, so evolution itself varies in direction, rate, and amount at different times.

6.3 Allometry

Haeckel's theory of recapitulation argued that evolution takes place by adding new stages at the end of normal development, so that an individual repeats the adult stages of its ancestors before reaching maturity (summarized by the aphorism that 'every species climbs its own phylogenetic tree'). This inhibited evolutionary biologists for over 50 years. It was finally disposed of by the recognition that although the early development of any group is similar in all members, divergence (and novelty) may occur at any point. However, the importance in evolution of changes in growth rate and body proportions (produced by differential growth rates or *allometry*) has been very great; perhaps the best known example is the enormous antlers of the extinct Irish elk, which are the 'right' size for an animal whose body size also has increased over that of its ancestors. The possibilities of a major change in proportions from a fairly small change in growth rate have been extensively documented by morphologists, especially D'Arcy Thompson and Julian Huxley. In general, developmental events can either be accelerated (i.e. appear earlier in ontogeny) or retarded (appear later). If sexual maturity appears later, then all other characters appear to have accelerated (since they appear earlier in relation to sexual maturity (and *vice versa*). *Neoteny* is the most commonly described case of a developmental shift, involving hitherto juvenile features appearing in the adult (as in the axolotl).

Neoteny is evolutionarily important because it offers an escape from specialization which does not involve a laborious assembly of a novel set of correlated genes. As far as major evolutionary steps are concerned, it is inconceivable that an adult invertebrate should become a vertebrate, or an adult coelenterate an annelid, *but* the transition may occur in similar embryos despite highly divergent adults; the embryos of some echinoderms and some protovertebrates are nearly identical.

Novelty can also appear as body proportions change; flower form depends on the pattern of fusion among parts, and in turn on developmental timing; a bird has fewer bones than a reptile due to fusions between primordia; simple changes in competitive interactions between cells can profoundly affect the number (and hence arrangement) of repeated parts (such as petals, digits, or scales), including pigment patterns; one of the best attested neotenous changes is the slowing down of human development* when compared with that in the great apes, so that the flat face, high relative brain weight and pelvic anatomy of the juvenile ape have become normal features of adult men and women. Such changes in growth rates can be based on very few gene differences; the oft-repeated fact that men and chimpanzees differ only in 1% of their DNA sequences may be a measure of the major effect of growth-regulating genes.

6.4 Macroevolution

Can the whole of evolutionary change be explained in terms of the same genetical forces that affect and produce species, or is it necessary to invoke additional mechanisms? The short answer to this question is that present and past diversity *can* be accounted for by known agents for maintaining and assorting variation (of which by far and away the most important is natural selection) (section 2.2). However it is impossible to disprove if other factors have

---

*The order of development is the same in mouse and man, but early stages develop two to four times slower in man than in mouse, and later stages five to 15 times.

ever acted. For example, the amniote vertebrates have a development which is like that of the anamniotes turned inside out. In a situation like this, it is always possible to argue that a mutation produced a 'hopeful monster' which may rarely and fortuitously survive to break through into a new habitat. In fact such a 'hopeful monster' can be regarded as a random genetic event in the same way that a random group of individuals may found a new population, and differ from the ancestral group (sections 2.2.1 and 6.2.2). Sooner or later all gene combinations are tested by natural selection; in the long term evolution is adaptive, but at any one moment, many new and untried genotypes will be exposed in both old and new environments. One of the recent advances in evolutionary understanding has been a better integration of ecological processes with genetical forces, including a recognition that all populations are not desperately struggling for existence all the time; there are occasions when 'hopeful monsters' can persist for finite periods, and perhaps find a niche for themselves. An old debate about whether all evolution is adaptive or not, is now falling into perspective.

One aspect of evolution which has not so far been considered is extinction. Extinctions are controlled by selection; which species is going to be adapted to a new environmental change cannot be predicted, but nonetheless is part of the evolutionary process. The rate of extinctions has varied considerable at different times in the fossil record, showing adaptational crises, particularly when whole families or classes are wiped out. The interesting point is that new higher taxa tend to appear after waves of extinction, suggesting that they are particularly dependent on the presence of unoccupied habitats available for colonization (Fig. 6–2). This would tend to confirm that evolution normally proceeds by adaptation in the usual way, rather than through hopeful monster formation (which might occur at any time, and not be dependent on previous extinctions).

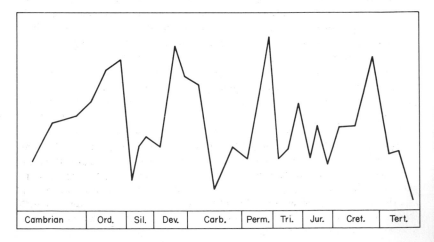

Fig. 6–2   Extinctions of shallow sea invertebrate families over the past 570 million years. (From Ayala and Valentine, 1979, *Evolving*. Benjamin Cummins, 1979.)

6.5 The evidence for evolution

We have already noted that Darwin convinced those of his colleagues prepared to accept evolution by putting forward a believable mechanism for change as well as by the weight of evidence (section 1.2). We have further seen that the evolutionary syntheses of the 1930s and 1970s depended mainly upon the interpretation of known data rather than the discovery of new facts. Indeed, it is intriguing how little the actual evidence for evolution has featured in the recurring controversies since 1858. Nevertheless it is obviously necessary to ask if the evolutionary interpretations described so far are based on real evidence of evolutionary change.

The fossil record in 1858 was poorly known, but the basic outline of the succession of life was well established. Darwin knew that trilobites were confined to the Palaeozoic and dinosaurs to the Mesozic, and many other features of the history of life. When palaeontologists had their eyes opened by Darwin, they turned the known but vaguely defined succession of life in the fossil strata into phylogenies; the classification of animals and plants was reorganized to reflect their evolution, and so-called phylogenetic or natural classifications were proposed for many groups.

Darwin sought and largely failed to find evidences of gradual change in fossils, and missing links between groups. He blamed the incompleteness of the fossil record. Several hundred 'missing links' are now known, but it is still true that we have more gaps than transitional forms. However, we now interpret this as the consequence of new species often arising from small populations in isolated areas (section 6.2.2), which is not unlike the answer Darwin himself gave to the problem (section 1.3.1). To some extent, it is possible for palaeontologists to estimate the degree of incompleteness of the fossil record; when this is done, it is clear that we have an adequate record of most preservable groups and that this supports phylogenies and replacements built from molecular, morphological, and biogeographical data. Nevertheless the record has to be read, and it is susceptible to an interpretation based on work with living populations as far as the nature and speed of particular evolutionary events are concerned (see section 6.2.2).

Standard textbooks on evolution devote much space to a discussion of 'homologies' (including vestigial organs), i.e. organs or processes in different groups which have a common origin. Common examples are the constancy of the DNA code, animal cell membranes, the structure of such macro-molecules as haemoglobin, mouthparts of insects, planktonic larvae of marine invertebrates, pentadactyl limb of vertebrates, flower parts of angiosperms, and the gametophyte stage of land plants. These all provide strong circumstantial evidence for evolution, and in toto as convincing a case as any of Sherlock Holmes's. But morphological homologies by themselves have never been very convincing. The problem is that morphology can often be explained by more than one hypothesis; different workers may differ about particular relationships and therefore weaken their use as evolutionary evidence. For example, the structures of lung fish have been a recurring subject for disagreement. However, the enormous mass of evidence from comparative morphology, physiology, and biochemistry far outweighs disputes over particular examples.

Darwin himself was convinced that evolution had occurred by his observations of geographical variation and replacement on the *Beagle* voyage. During the 1920s there was a belief that geographical variation was a phenotypic response to different environments. This was shown to be untrue by Turesson's demonstration that ecotypic variation in plants is inherited, and Sumner's similar work on deermice (*Peromyscus* species). Biogeographical data (supplemented by geological information on continental drift, sea-level changes, etc. and ecological information on competition, dispersal, tolerance, etc.) provide strong support for evolutionary change.

Sometimes 'facts' are reported that 'disprove' evolution (such as the alleged occurrence together of fossil human and dinosaur footprints). Such apparent anomalies obviously need examining on their merits, but it is worth emphasizing again that the strength of the neo-Darwinian understanding is its proven mechanism rather than any accumulation of evidence. Indeed, one of the more serious complaints about neo-Darwinism is that it explains so much that it has become dogma rather than science, and should be rejected on that ground. This is a criticism we shall examine in the final chapter (section 7.1).

7 Was Darwin Wrong After All?

Darwinism has been under attack ever since the *Origin* was published, but the objections raised in the 1860s, 1900s, 1930s and 1960s have all concluded with a strengthening of Darwin's original propositions. A major weakness in Darwin's own understanding of the natural world was resolved with the re-discovery of Mendel's work, and the integration of particulate inheritance into evolutionary theory. This confirmed rather than changed the arguments of the *Origin*. Weismann's theory of germ-line separation and early interpretations of the randomness of mutation were both important in the development of evolutionary thinking (*q.v.* sections 1.3 and 1.5.1), and both have had to be modified (the former by the recognition that it only applies to higher animals; the latter by the discovery that much mutation is under biochemical control, and not by chance in the statistical sense). Neither affects the neo-Darwinian understanding of biological change as developed in the first six chapters of this book.

Despite its survival so far, neo-Darwinism is still being attacked. In the 1980s there is no major objection, but at least five different series of criticisms. In this chapter, these five criticisms are examined, and their significance assessed.

7.1 Philosophy

Karl Popper has argued that neo-Darwinian theory cannot be falsified, and therefore that it is metaphysics and not science. Other critics have argued that natural selection is tautologous, since the survivors are the fittest – and these are the survivors. Neither of these objections stands up to detailed investigation: the ingredients of neo-Darwinian theory (natural selection, competition, speciation mechanisms, etc.) are all subject to experimental test in normal rigorous ways, even if the whole neo-Darwinian edifice (or *paradigm*) cannot be so examined; and natural selection is not tautologous if 'fitness' is correctly interpreted in its biological sense (as a measure of reproductive success) instead of its common usage as meaning health or fitness, and even less so when Darwin's original insistence that fitness is a variable is included.

The philosophical problem seems to boil down to the fact that evolutionary theory is by its nature synthetic, and bears the hallmark of any good science in uniting many areas of knowledge into one integrated whole. Within this whole it is relatively easy to make falsifiable claims, but their proof or disproof will not affect the main features of the synthesis. For example, much of evolution is adaptive; it will not falsify the neo-Darwinian theory to show that some non-adaptive evolution occurs. Indeed, it is of considerable interest to know how much non-adaptive evolution occurs. It may be that neo-Darwinism falls outside a particular definition of science, but it is not true that it is therefore not science; bringing together different (and testable) ideas from a wide range of disciplines must surely be a scientific activity. Indeed Popper has commented (1980) on his alleged views, that 'some people think I have denied scientific character to the historical sciences, such as palaeontology, or the history of the evolution of life on Earth. This is a mistake, and I wish to affirm that these and other historical sciences have in my opinion scientific character; their hypotheses can in many cases be tested'.

However, evolutionary biologists probably have a worse record for speculation than virtually any other scientists. Speculation is valid if it is used to form a hypothesis which can be tested, but far too much evolutionary speculation has remained at the level of guess-work. In the nineteenth century the worst offenders in this respect were comparative morphologists; now they tend to be behavioural biologists (or sociobiologists). More dangerous have been ethicists, economists, eugenicists, and other non-biologists who have used analogies from evolutionary biology in their own disciplines. This is certainly not science.

The philosophers of science have performed a useful function in forcing practising biologists to examine their science for woolliness and false assumptions, but on balance their criticisms have probably done more harm than good by leading outsiders into thinking that evolution is closer to astrology or alchemy than respectable biology.

7.2   Punctuated equilibria

In 1972, two American palaeontologists, Nils Eldredge and Stephen Jay Gould pointed out that the most marked characteristic of the fossil record was not long-continued change of particular forms, but periods of stability (or stasis), followed by the sudden appearance of a new (albeit related) form. In effect, they challenged the prevailing orthodoxy that our lack of knowledge of the origin of fossil species was the result of gaps in the record, and claimed that the unbiased interpretation of fossil successions implies that evolution proceeds by fits and starts. This has been taken by some as an argument for instant (or quantum) speciation, produced by major mutations in the sense of de Vries (section 1.5). In fact Eldredge and Gould claimed nothing of the sort: all their emphasis means is that evolutionary rates vary considerably, and that the speciation process is comparatively rapid. There is no problem about this; as far as speciation is concerned, this is expected from a consideration of speciation mechanisms (section 6.2.2).

Punctuated equilibria have been contrasted with Darwin's own 'gradualism'. This is a false opposition. Darwin argued against evolutionary 'jumps' or saltations, and no convincing evidence for such has ever been produced. The idea of punctuated equilibrium is a valuable recall to the realities of the fossil record, and to the importance of stabilizing selection and rapid speciation as genetic forces.

7.3   Cladism

Taxonomy has traditionally been a subjective exercise, controlled by the general agreement between specialists as to the reality of particular species. For many years biologists have tried to make classification more exact. This has involved the consideration of a greater number of characters (including chemical and behavioural) than the usual morphological criteria, and, increasingly, the use of quantitative multivariate techniques. One particular refinement of this trend is cladism, a method of systematics derived from the work of a German taxonomist Willi Hennig (and based on a book by him published in English in 1966). Cladism has three axioms:

1) Features shared by organisms manifest a hierarchical pattern in nature

2) This hierarchical pattern is economically expressed in branching diagrams (cladograms)

3) The nodes (branching points in a cladogram) symbolize the homologies shared by the organisms linked by the node, so a cladogram is synonymous with a classification.

The method imposes a discipline on taxonomists, but in the process has become divorced from evolutionary biology, and especially the working hypothesis of most systematists that they are largely reconstructing phylogeny in their classifications. Indeed some cladists explicitly disavow that they are concerned with evolution; they claim that they are concerned solely with 'pattern' in nature, in ways reminiscent of pre-Darwinian romantics such as Goethe and Geoffrey St Hilaire. In this respect they have departed from Hennig's own method, which was consciously phylogenetic; ironically the major theoretical contribution of Hennig to classification was in distinguishing between primitive and derived homologies, which is necessary before a phylogeny can be constructed.

The cladism controversy has obscured a fundamental principle that there can be only one true phylogeny for a group, but there cannot be a simple classification in the same sense, because the rules of classification are made by people for their own convenience. It has achieved an undue notoriety in Britain, because of its advocacy by some members of the British Museum (Natural History) staff. The best way to regard cladism is simply as a valiant attempt by professional systematists to improve their particular technology. It neither aids nor hinders evolutionary biology.

7.4 Development

The doubts raised by embryologists about Darwinism are the least convincing of the 1980s criticisms. This is not because the problems concerned are trivial or irrelevant, but because of the apparent difficulty of posing them in modern times. This may be because advances in molecular biology have enabled many of the basic problems in mutation and variation to be set out in very clear-cut terms, but many embryological problems about the origin of pattern and the so-called 'laws of form' are still expressed in the almost mystical terms of 'morphogenetic fields' used around the turn of the century. Few biologists have attempted to analyse development genetically (notable exceptions are C. H. Waddington, E. Hadorn, and H. Grüneberg). However molecular biologists have now begun to turn their attention to the effects of gene substitution on organ structure and behaviour. It is clear from the comparatively small amount of work that has been done that gene action early in development may profoundly modify the form and function of the adult (see section 6.3), and

Waddington (in particular) has shown how genes of small effect can be selected by stresses imposed through the environmental modification of development. There are many links needing to be forged between evolutionary biology and development; it is at best premature to regard embryological criticisms of neo-Darwinism as serious.

7.5   Creationism

The most persistent challenge to neo-Darwinism comes from certain Christian groups, and also some Muslim sects. Although the recognition that the world was many thousand years old grew during the first half of the eighteenth century and Christian opinion was introduced to the idea of evolution by such books as Robert Chambers's *Vestiges of Creation* (1834), until the *Origin* most people still thought of the world and all its species as being designed and made by God the Great Watchmaker in October 4004 BC (Archbishop Ussher's calculation, made from adding up the ages of the Old Testament patriarchs).

Darwin's argument that the present state of biological diversity has arisen by an entirely mechanistic process was anathema to many and has remained so. The prosecution in 1925 of John Scopes for teaching Darwinism in a Tennessee school is well-known. For forty years after that, little was heard of the creationist controversy. It surfaced again with a requirement by the California State Board of Education (1972) that creation be accorded 'equal time' with evolution in schools within the state. Implications and challenges of this ruling continued ever since, and are increasingly being felt in Britain.

The problem about evolution for 'creationists' is that it seems to make man merely a monkey; the best-known expression of this is Benjamin Disraeli's comment 'Is man an ape or an angel? . . . I am on the side of the angels. I repudiate with indignation and abhorrence these new fangled theories'. This is not the place to review the religious debates about the relation of man to the rest of the world. However it is pertinent to note that questions about God and the spiritual nature of man are not scientific ones; indeed they are outside the realm of science. T. H. Huxley was very condemnatory about the influence of religion on science, but he was realistic about the nature of man: 'It is the secret of the superiority of the best theological teachers to the majority of their opponents, that they substantially recognize the realities of things. The doctrines of predestination; of original sin; of the innate depravity of man appear to me to be vastly nearer the truth than the 'liberal' popular illusions that babies are all born good and that the example of a corrupt society is responsible for their failure to remain so; and that it is given to everybody to reach the ethical ideal if only he will try'.

Creationists insist that the physical history of the human race is identical with God's creation of it, and extrapolate from this to deny that any significant evolution has occurred. Many of the disputes recorded in this book are taken out of their contexts and used by creationists as support for their contention that evolution and evolutionary theory is nothing more than dubious hypothesis. For example, it is claimed that all the phyla except the vertebrates are present in the earliest fossiliferous rocks (ignoring Precambrian fossils), that the theory of punctuated equilibria is inconsistent with neo-Darwinism, that all mutations are harmful, that the most important hereditary determinants are in cell membranes and not nuclear DNA, that no species transitions are known (except the trivial case of polyploidy), that evolution is contrary to the Second Law of Thermodynamics (which only applies in a closed system, but the earth is continually receiving solar energy), that all rock-dating techniques are unreliable, and so on. Creationist literature is a sadly distorted microcosm of the whole neo-Darwinian debate. The tragedy of the creationist position is that it begins from the axiom that evolution is an impossibility, and hence preempts rational discussion on the issue.

7.6   The fifth synthesis?

The common characteristic of all the attacks and debates about evolution, past and current, is an over-emphasis on one particular factor or a failure to take into account data from disparate disciplines. Difficulties in this respect increase as biological knowledge grows, and it becomes difficult to keep different parts of the subject in focus. One reason why the history of evolutionary debate is important is that old arguments tend to recur. None of the objections recorded in this chapter is new, and much time and effort would have been saved if the protagonists had been better versed in past debates. Indeed one may honestly (if rashly) assert that there is now no real attack on the neo-Darwinian synthesis as it has developed over more than 100 years. Work on punctuated equilibrium should give further insight into speciation processes, whether or not Eldridge and Gould are confirmed in all their beliefs; there is a need for more information about the importance of embryological patterns in evolution, but it is difficult to see how it may lead to any major adjustment in the processes described in Chapters 5 and 6; and as for cladists and creationists, they are trying to fight battles in areas where the war has long passed.

It is dangerous to suggest that any major set of scientific ideas is close to its final form; as far as the neo-Darwinism is concerned, the neutralist debates of the late 1960s and 1970s are still too close to allow complacency (particularly while creationists allege that evolutionists have been brain-washed). Nevertheless it seems fair to claim that the neo-Darwinian synthesis is in good shape with new insights from the neutralist debates on the maintenance of variation and the interaction of genes with ecology; and that there is no need for any major new synthesis to incorporate the attacks of the 1980s.

8 Envoi

This account of neo-Darwinism began with closet biologists, trying to interpret the whole of biology around their own speciality. Repeatedly we have returned to the problems that have arisen through intellectual narrowness. However, one strand of biology has always acted as a corrective to this narrowness, and that is the long and peculiarly British tradition of the natural historian, which inevitably maintains the unity of biology at times when geologists, biochemists, geneticists, theoreticians, and others are tending to rush off in different directions.

It is import that all biologists should be to some extent natural historians, since this would keep them in touch with biological reality. Darwin was a naturalist; so were Wallace, Bates, Weldon; so in a different way were R. A. Fisher and J. B. S. Haldane. Unfortunately natural history is not an academic discipline; it is a commitment which is caught much like a bacterium acquiring a favourable plasmid.

Darwin wrote in his *Autobiography* that his father had once accused him 'You care for nothing, but shooting, dogs, and rat-catching' and he claimed of himself that 'nothing could have been worse for the development of my mind than Dr Butler's school, as it was strictly classical, nothing else being taught, except a little ancient geography and history . . .'. This is not the place to pass judgment on Darwin's or anyone else's eduction, but it is pertinent to repeat once again that virtually all the arguments about evolution have been put forward by people unable to appreciate the breadth of the synthesis achieved by Darwin, a synthesis built upon as biology *sensu stricto* has advanced. We have here a proper application of Pope's often ill-applied exhortation, 'A little learning is a dang'rous thing, Drink deep or taste not the Pierian spring: There shallow draughts intoxicate the brain, And drinking largely sobers us again'.

However, it is fitting to let Darwin have the last word; his description of himself will encourage some, but I hope stimulate all. 'I have no great quickness of apprehension or wit which is so remarkable in some clever men, for instance, Huxley. I am therefore a poor critic; a paper or a book, when first read, generally excites my admiration, and it is only after considerable reflection that I perceive the weak points. My power to follow a long and purely abstract train of thought is very limited; and therefore I could never succeeded with metaphysics or mathematics. My memory is extensive, yet hazy; it suffices to make me cautious by vaguely telling me that I have observed or read something opposed to the conclusion which I am drawing, or on the other hand in favour of it; and after a time I can generally recollect where to search for my authority. So poor in one sense is my memory, that I have never been able to remember for more than a few days a single line of poetry.'

'On the favourable side of the balance, I think that I am superior to the common run of men in noticing things which easily escape attention, and in observing them carefully. My industry has been as great as it could have been in the observation and collection of facts. What is far more important, my love of natural science has been steady and devout'.

Further Reading

A proper documentation of all the issues discussed in this volume would occupy many pages. It seems more profitable simply to list some of the more useful books for further reading.

General introductions and elementary textbooks

BERRY, R. J.(1977). *Inheritance and Natural History*, 350p. Collins New Naturalist, London.

CAIN, A. J. (1954). *Animal Species and their Evolution*, 190p. Hutchinson's University Library, London.

DOWDESWELL, W. H. (4th ed. 1975). *The Mechanism of Evolution*, 156p. Heinemann Scholarship Series, London.

GRANT, V. (1963). *The Origin of Adaptations*. 606p. Columbia U.P., London & New York.

GRANT, V. (1977). *Organismic Evolution*, 418p. Freeman, San Francisco.

MAYNARD SMITH, J.(3rd ed. 1975). *The Theory of Evolution*, 344p. Pelican, London.

PARKIN, D. T. (1979). *An Introduction to Evolutionary Genetics*, 233p. Edward Arnold, London.

PATTERSON, C. (1978). *Evolution*, 197p. Routledge & Kegan Paul, and British Museum (Natural History), London.

PAUL, C. (1980). *The Natural History of Fossils*, 292p. Weidenfeld & Nicolson, London.

RHODES, F. H. T. (1962). *The Evolution of Life*, 302p. Pelican, London.

WILSON, E. O. and BOSSERT, W. H. (1971). *Primer of Population Biology*, 192p. Sinauer, Stamford, Conn.

Advanced text-books

AYALA, F. J. (ed.) (1976). *Molecular Evolution*, 277p. Sinauer, Sunderland, Mass.

BROWN, J. L. (1975). *The Evolution of Behaviour*, 761p. Norton, New York.

DOBZHANSKY, T., AYALA, F. J., STEBBINS, G. L. and VALENTINE, J. W. (1977). *Evolution*, 572p. Freeman, San Francisco.

ENDLER, J. A. (1977). *Geographic Variation, Speciation and Clines*, 246p. Princeton U.P., Princeton, N.J.

FOREY, P. L. (ed.) (1981). *The Evolving Biosphere*, 311p. Cambridge U.P. and British Museum (Natural History), London.

FUTUYMA, D. J. (1979). *Evolutionary Biology*, 565p. Sinauer, Sunderland, Mass.

GOULD, S. J. (1977). *Ontogeny and Phylogeny*, 501p. Harvard U.P., Cambridge, Mass. & London.

LEWONTIN, R. C.(1974). *The Genetic Basis of Evolutionary Change*, 346p. Columbia U.P., New York.

OPEN UNIVERSITY(1981). *Course S364 Evolution* (15 units). Open U.P., Milton Keynes.

STANLEY, S. M.(1979). *Macroevolution, Pattern and Process*, 332p. Freeman, San Francisco.

WADDINGTON, C. H. (1957). *The Strategy of the Genes*, 262p. Allen & Unwin, London.

WHITE, M. J. D. (1978). *Modes of Speciation*, 455p. Freeman, San Francisco.

WILLIAMSON, M. (1981). *Island Populations*, 286p. Oxford U.P., Oxford.

Classical works

Foremost in this list must be Charles Darwin's own books, notably *the Origin of Species* (available in many editions).

DE BEER, G. R. (1930). *Embryology and Evolution*, 116p. Clarendon. Oxford (revised as *Embryos and Ancestors* (1940, 3rd ed. 1958)).

DOBZHANSKY, T. (1937, 3rd ed. 1951). *Genetics and the Origin of Species*, Columbia U.P., New York (revised as *Genetics of the Evolutionary Process* (1970)).

FISHER, R. A. (1930, 2nd ed. 1958). *The Genetical Theory of Natural Selection*, 272p. Clarendon, Oxford (2nd ed. published by Dover, New York).

FORD, E. B. (1931). *Mendelism and Evolution*, Methuen, London.

HALDANE, J. B. S. (1932). *The Causes of Evolution*, 235p. Longmans, London and Harper, New York.

HUXLEY, J. S. (1932). *Problems of Relative Growth*, 276p. Clarendon, Oxford.

HUXLEY, J. S. (1942). *Evolution, The Modern Synthesis*, 645p. Allen & Unwin, London.

MAYR, E. (1942). *Systematics and the Origin of Species*, 334p. Columbia U.P., New York (revised as *Animal Species and Evolution (1963)*, published by Harvard U.P., Cambridge, Mass. & Oxford. U.P., London).

SIMPSON. G. G. (1944). *Tempo and Mode in evolution*, 237p. Columbia U.P., New York (revised as *The Major Features of Evolution* (1953)).

STEBBINS, G. L. (1950). *Variation and Evolution in Plants*, 643p. Columbia U.P., New York.

Histories of evolution

MAYR, E. and PROVINE, W. B. (eds) (1980). *The Evolutionary Synthesis*, 487p. Harvard U.P., Cambridge, Mass. & London.

MOORE, J. R. (1979). *The Post-Darwinian Controversies*, 502p. Cambridge U.P., London.

OSPOVAT, D. (1981). *The Development of Darwin's Theory*, 301p. Cambridge U.P., Cambridge.

PROVINE, W. B. (1971). *The Origins of Theoretical Population Genetics*, 201p. Chicago U.P., Chicago & London.

There are biographies or autobiographies of many of the chief actors in the neo-Darwinian saga, including Darwin and Wallace; also W. Bateson, R. A. Fisher, Francis Galton, J. B. S. Haldane, T. H. and J. S. Huxley, Gregor Mendel, and T. H. Morgan.

Critics of neo-Darwinism (This is a very selective list)

CANNON, H. G. (1958). *The Evolution of Living Things*, 180p. University Press, Manchester.

ENOCH, H. (1967). *Evolution or Creation*, 172p. Evangelical, London.

GRASSÉ, P.-P. (1977). *Evolution of Living Organisms*, 297p. Academic Press, New York & London.

HOYLE, F. and WICKRAMSINGHE, (1981). *Evolution from Space*, 176p. Dent, London.

KERKUT, G. A. (1960). *Implications of Evolution*, 174p. Pergamon, Oxford.

MACBETH, N. (1974). *Darwin Retried*, 178p. Garnstone, London.

MOOREHEAD, P. S. and KAPLAN, M. M. (1967). *Mathematical Challenges to the Neo-Darwinian Interpretation of Evolution*, 140p. Wistar Institute, Philadelphia.

WHITCOMB, J. C. and MORRIS, H. M. (1961). *The Genesis Flood*, 518p. Presbyterian & Reformed, Philadelphia.

Index